Book Description

There is a hidden agenda in Elon Musk and Jeff Bezos's space exploration to colonize Mars and the Moon. The media portray them as Americans who are rekindling space exploration that was once dominated by the rivalry between the United States and the extinct Soviet Union.

Before you write off what I have just said, hear me out. For more than a decade, I've closely studied the lives of Bezos and Musk. I have watched as Jeff Bezos dominated online retailing, and Elon Musk building Tesla and SpaceX. As each spread their wings to critical industries, I began to notice a pattern that only leads to one conclusion, the hidden agenda each harbors.

On the surface, it appears each of them is interested in dominating space. However, when you study how they operate in every industry, you cannot help but conclude that there is more to their modus operandi.

If you would like to find out what this hidden agenda is, and how Musk and Bezos are dominating the industries they each operate in, then *BEZOS vs. MUSK* is a book you should read.

In this book, I, Dan Bleigh, an MBA graduate, will be providing a unique angle of looking into the innovations and technology that Bezos and Musk use to dominate. I compare the two top richest people in the world side by side. You'll learn the logistics strategy that each of these entrepreneurs applies to move their companies to the top.

Here's more of what you'll learn in this book:

- **Two dreams that Elon Musk has that'll take him to the top of the air transportation industry.**
- Discover two images in the conclusion that shows what Elon Musk and Jeff Bezos are planning when they're done with their respective projects.
- What Jeff Bezos did to force NASA to freeze the Artemis program for a few months.
- An initiative that Elon Musk is undertaking to help increase the thinking power of human beings and why he's doing it.
- Why Elon Musk and Amazon Web Services have funded OpenAI.
- **A potent strategy that Amazon uses to dominate an industry it never operated in before.**

And much much more!

The rivalry between Jeff Bezos and Elon Musk is intense, and these two billionaires are in conflict over NASA's contracts. However, these conflicts are hiding what is happening in reality.

If you want to find out the truth behind the rivalry of Bezos and Musk, get *BEZOS vs. MUSK*.

BEZOS vs. MUSK

A GLOBAL BATTLE

Dan Bleigh

© **Copyright 2021 - All rights reserved.**

The content contained within this book may not be reproduced, duplicated or transmitted without direct written permission from the author or the publisher.

Under no circumstances will any blame or legal responsibility be held against the publisher, or author, for any damages, reparation, or monetary loss due to the information contained within this book, either directly or indirectly.

Legal Notice:

This book is copyright protected. It is only for personal use. You cannot amend, distribute, sell, use, quote or paraphrase any part, or the content within this book, without the consent of the author or publisher.

Disclaimer Notice:

Please note the information contained within this document is for educational and entertainment purposes only. All effort has been executed to present accurate, up to date, reliable, complete information. No warranties of any kind are declared or implied. Readers acknowledge that the author is not engaged in the rendering of legal, financial, medical or professional advice. The content within this book has been derived from various sources. Please consult a licensed professional before attempting any techniques outlined in this book.

By reading this document, the reader agrees that under no circumstances is the author responsible for any losses, direct or indirect, that are incurred as a result of the use of the information contained within this document, including, but not limited to, errors, omissions, or inaccuracies.

Cover Photos Credits

Note: Both images are free to use under creative commons license

Jeff Bezos' image: Oberhaus, D. (2019, May 9). Jeff Bezos. Flickr. https://www.flickr.com/photos/163370954@N08/32878819397

Elon Musk's image: Jurvetson, S. (2019, April 22). Elon Musk presenting Tesla's fully autonomous future. Flickr. https://www.flickr.com/photos/jurvetson/40705940233/in/photostream/

Table of Contents

INTRODUCTION 1

CHAPTER 1: THE SIZE OF GLOBAL LOGISTICS, AND AN INTRODUCTION TO BEZOS' AND MUSK'S COMPANIES AND PROJECTS AIMED AT DOMINATING GLOBAL LOGISTICS 9

    REVEALED: THE HISTORY OF LOGISTICS AND SUPPLY CHAIN 10
    AMAZON: WHAT THIS U.S. COMPANY IS REALLY ABOUT 12
    AMAZON'S LOGISTICS AND DELIVERY STRATEGY 16
        What Is Amazon Air? 17
        Amazon's Ground Transportation Strategy 18
        Other Amazon Projects You Should Know About 20
    ELON MUSK'S DISRUPTING THESE INDUSTRIES 22
        The Automotive Industry 23
        SpaceX, Elon Musk, and the Telecommunications Industry 27
        Elon Musk and His Revolution of the Energy Sector 29
        Elon Musk Introduces a New Mode of Human Transport 31
        How Elon Musk Joined the Tunneling Industry 34
        Why Musk Is Interested in Artificial Intelligence (AI) 36
        What's Elon Musk Doing in Health Care? 38

CHAPTER 2: THE TIP OF THE ICEBERG: THE SPACE RACE 41

    ELON MUSK, SPACEX, AND AEROSPACE 42
    JEFF BEZOS, BLUE ORIGIN, AND AEROSPACE 45
    THE MUSK-BEZOS SPACE RIVALRY 46
    SPACEX WINS NASA CONTRACT TO THE MOON 47
    BLUE ORIGIN'S FIRST PASSENGER SPACEFLIGHT 48
    ELON MUSK SENDS ASTRONAUTS TO THE INTERNATIONAL SPACE STATION (ISS) 49

CHAPTER 3: WHO WILL DOMINATE THE LOGISTICS RACE BETWEEN MUSK AND BEZOS? 51

INTRODUCTION TO AMAZON ROBOTICS 52
WHAT YOU SHOULD KNOW ABOUT AMAZON LOGISTICS 54
   Why Amazon Went Into Product Shipping 55
BLUE ORIGIN AND AIRSPACE LOGISTICS 57
THE BIRTH OF TESLA LOGISTICS 57
   Tesla Factories You Should Know About 60
PRODUCTION AND LOGISTICS OF SPACEX ROCKETS 62

CHAPTER 4: MUSK AND BEZOS IN THE RACE TO DOMINATE ARTIFICIAL INTELLIGENCE (AI) 65

MUSK AND ARTIFICIAL INTELLIGENCE IN HIS COMPANIES 66
   What Tesla Artificial Intelligence Is About 66
   What Is Neuralink Artificial Intelligence? 69
OPENAI AND ITS ROLE IN ARTIFICIAL INTELLIGENCE 71
BEZOS AND AMAZON'S ARTIFICIAL INTELLIGENCE 73
   The Root of Amazon's Machine Learning and AI 74

CHAPTER 5: WILL BEZOS BEAT MUSK IN THE SELF-DRIVING VEHICLES RACE? 79

WHY TESLA COULD WIN THE SELF-DRIVING VEHICLES RACE 80
   How Tesla Plans to Solve the Shortage of Trucks in the United States 83
BEZOS TAKES ON TESLA WITH AMAZON ZOOX 84
DISCOVER WHY AMAZON INVESTED IN AURORA 87

CHAPTER 6: BEZOS BATTLES MUSK IN THE ELECTRIC VEHICLES RACE; CAN HE WIN THIS RACE? 91

WHAT IS AMAZON RIVIAN? 92
AMAZON ELECTRIC TRUCK FOR THE MIDDLE MILE 95
TESLA: THE TOP DOG OF ELECTRIC CAR MANUFACTURERS 97
   Tesla's Upcoming Electric Pickup: The Cybertruck 98
   Is the Tesla Van in the Cards? 99
   The Boring Company's Tunnel to California Airport and Electric Vans 100

CHAPTER 7: HOW SPACEX PLANS TO TRANSPORT CARGO BY AIR AND WHAT BEZOS IS DOING ABOUT IT 103

SPACEX CONTINENT-TO-CONTINENT ROCKET CARGO TRANSPORTATION
....104
MUSK DREAMING ABOUT A SUPERSONIC JET................................ 105
WHAT AMAZON IS DOING IN THE AIR-TRANSPORTATION SPACE 107

CHAPTER 8: WHO WILL WIN THE TIGHTLY CONTESTED FREIGHT-TRANSPORTATION RACE? 113

AMAZON EXTENDS ITS TENTACLES INTO FREIGHT TRANSPORTATION . 114
AMAZON IN THE FREIGHT BROKERAGE SPACE 115
INTRODUCES THE TESLA SEMI FOR FREIGHT TRANSPORTATION 117
THE BORING COMPANY DIGS FREIGHT-TRANSPORTATION TUNNELS.. 119
INTER-CITY HYPERLOOP CONTAINER TRANSPORT 122

CHAPTER 9: THE RACE TO DOMINATE GLOBAL SATELLITE-BASED BROADBAND INTERNET SERVICE 125

WHY THERE'S NEED FOR SATELLITE INTERNET............................... 126
THE CLASH OF AMAZON'S PROJECT KUIPER AND SPACEX'S STARLINK 127
PROGRESS OF PROJECT KUIPER 129
ELON MUSK WANTS TO CONNECT STARLINK SATELLITE INTERNET TO
MOVING VEHICLES 130
Self-Driving Cars and Starlink 132

CONCLUSION.. 137

REFERENCES... 145

Introduction

Jeff Bezos and Elon Musk are two of the richest people in the world. Whatever they do quickly hits the media because the audience jumps at such news quickly. The problem with media is that it focuses on hype, trends, and rivalries between public figures to attract audiences to their news. The same is happening with the rivalry between Jeff Bezos and Elon Musk regarding their space projects, Blue Origin and SpaceX, respectively. On the surface, it appears the Bezos and Musk rivalry is about dominating space; however, the hidden reality is that each wants to dominate Earth's global logistics. Anyone who hasn't looked at the strategies from these two wealthy entrepreneurs may not discover what I've just said. That's partly why I've written this book—for a person like you to understand the real battle going on between Jeff Bezos and Elon Musk.

The space race has just gained further momentum in early 2021 thanks to the richest men in the world, Jeff Bezos and Elon Musk. The new conquest of space, this time not between two great world powers, like the U.S. and the former USSR in the 1950s until the fall of the Communist dictatorship at the end of 1991, but between two American entrepreneurs who aim to colonize space shortly. In reality, SpaceX and Blue Origin's megaprojects to create future space colonies on the planet Mars or the moon, respectively, are just the tip of the iceberg, hiding the real race to conquer global logistics on Earth in the short run.

In this book, I'm going to help you understand the different links in the global logistical chain of Bezos and Musk by comparing them side by side, and see the extent of the efforts made by them to cut across the realm of commercial transport with breakthrough, innovation, and cutting-edge strategies and technology within less than a decade from now. These two billionaires are playing in various industries because they keep innovating—a key ingredient if you want to disrupt the world. Innovation doesn't mean creating new things, but using old things in new and enterprising ways. You'll discover how Musk and Bezos are doing this, throughout this book.

More importantly, the purpose of this book is to dissect the main delivery process of Amazon. On the side of Tesla, this book will bring together the missing elements, which seem to have no visible links, to build the blueprint for the creation of a logistics fleet empire based on the reduction of transport time by increasing delivery speed, energy efficiency, and self-driving vehicles using transport by relying on renewable energies.

Here's a detailed outline of what you'll learn in this book as I lay the path for you to see the hidden reality of global logistics domination by both Musk and Bezos. I begin the book by taking a step back and tracing the history of logistics and supply chain. This is important because it sets the foundation for what is to come throughout the book. You'll also learn about the state of global and U.S. logistics so that you can understand what Bezos and Musk stand to gain by dominating this industry.

The other important part of the foundation needed is to have a helicopter view of the industries Bezos and Musk are trying to dominate. For this reason, I give you a background on Amazon and where it stands in terms of its size. As much as it is a dominating online retailer, Amazon has one major challenge. This problem is part of the reason why Amazon is involved in many industries. However, each of those industries helps Amazon deliver on its promises to its customers. Musk, on the other hand, runs Tesla, one of the most valuable carmakers in the world. However, he's involved in at least six other industries, all of which you'll learn about in Chapter 1.

In Chapter 2, I'll dive deeper into the race for space travel that's currently going on between Bezos and Musk. In the 1950s up to 1991, the space race was dominated by nations such as the United States and the extinct Soviet Union. In 2003, the space race started to take a different turn when the Columbia shuttle broke into pieces as it returned to Earth from the International Space Station. In the process, the shuttle disaster killed seven crew members. This opened the way for private companies to be considered for space missions. At the time, Bezos and Musk had already started their respective space travel companies. Musk's SpaceX has dominated the space race by winning several NASA and military contracts. You'll learn about how Blue Origin and SpaceX are fighting over one contract, which has resulted in NASA freezing that contract for a few months.

The next race I will take you through is the logistics race. Amazon was founded 10 years before Musk joined Tesla as chairman in 2004. So, Amazon had a head start in developing its logistics network. As I'll show in Chapter 3, Amazon's logistics took off after it created a robotics and logistics arm. Furthermore, Amazon made strategic acquisitions that moved it quicker toward developing a giant logistics network. Amazon Robotics designs and builds robots that are used to improve operational efficiencies at Amazon's warehouses and sortation centers. On the logistics front, Amazon launched a shipping and delivery service that has jumped to number four in terms of size.

In contrast, Tesla had car production and logistics issues when it ramped up the production of its Model 3 sedan. These challenges, in hindsight, were good things that could have happened to Tesla. It had to innovate and come up with solutions, and it did by creating three solutions to their logistics issues. You'll learn what those solutions are in Chapter 3.

In Chapter 4, I shift attention to the world of artificial intelligence (AI). This is a field in which machines mimic human intelligence and help companies improve efficiencies and lower costs. Both Amazon and Tesla are big on artificial intelligence and machine learning. Tesla uses AI to develop self-driving electric vehicles. Because Tesla has more than one million cars on the roads globally, its engineers have a huge set of data to develop potent AI. Musk is applying AI in Neuralink, another of his companies, that operates in the health-care sector. This company has developed a chip, called Link, to help address brain-disorder issues, such as brain damage, seizures, and insomnia. The technology has been proven on pigs and should now be tried on people.

Amazon began using AI from 1999, mainly in its online shop. From that humble beginning, Amazon's AI has grown, and, today, this company boasts more than 200,000 robots at its sortation centers and warehouses. Furthermore, Amazon is using AI at its Amazon Go stores and Amazon Web Services (AWS).

The next race I look at is the race of the self-driving vehicles, and I go through it in Chapter 5. Amazon is not a carmaker and doesn't seem to be taking that path from scratch. This online retailer's strategy is to either buy self-driving vehicle-making companies or invest heavily in them. One thing is clear: Amazon is moving toward the usage of autonomous cars. There are two companies in which Amazon has invested in recently, and they make Amazon a part of the drive toward autonomous vehicles. One of the companies recently launched its first self-driving vehicle. You'll find the details in this chapter.

There is no company that has made a fully self-driving car, yet. Don't be fooled by Tesla when it says that it offers a fully self-driving option on its cars. Tesla itself knows it doesn't produce autonomous cars, as it says in its letter to California's Department of Motor Vehicles (DMV). However, Musk's company is hard at work developing such vehicles and is currently beta-testing its Full Self Driving (FSD), version 10.

In Chapter 6, I take you through the electric-vehicle race, an industry that Tesla undoubtedly dominates.

Tesla has been in this industry since 2003, for all its life. It has made and delivered more than one million electric vehicles and, in the process, has become one of the most valuable carmakers in the world. It differs from most other electric carmakers in that it tries to produce critical parts by itself. In other words, it strives toward vertical integration. One of the big challenges Tesla faces is low supply of batteries. Just as this vehicle-maker has done in the past, it is developing its own solution. The big question in many people's minds is whether Tesla will produce a van or not. You'll learn what my take and others' thoughts are on this subject.

Just as it does in the self-driving space, Amazon is not starting an electric-vehicle arm. Instead, it buys or invests in companies that allow it to benefit from the industry. Unsurprisingly, Amazon has invested in an electric-vehicle startup and also bought 100,000 vans from it. You'll learn the name of the company in Chapter 6. Furthermore, Amazon ordered electric trucks from a Canadian company about to list on a popular stock exchange. All these investments and acquisitions are helping Amazon move toward its goal of using 100% renewable resources by 2040.

Most people are familiar with air transportation in the form of ships and drones. What if I told you that you could travel from New York City to Shanghai in a rocket-propelled airship? Wouldn't you say I've lost my marbles? Most people would consider that a pipe dream, yet, Elon Musk has such a dream. I give you the details of that dream in Chapter 7 on the air-transportation race. SpaceX is developing a rocket called Starship, and Musk dreams of using this rocket to not only send people to the Moon and Mars, but also for city-to-city travel, continent-t0-continent people journeys, and continent-to-continent cargo transportation. He's also dreaming of making a superjet fueled by batteries. What's needed for a superjet is a powerful enough battery for take-off.

Amazon is already dominating the sky with aircraft and drones. This mega online retailer uses a fleet of aircraft to transport parcels from airport to airport. The ultimate aim is the timely delivery of customer packages. This will be supplemented by a drone delivery service, which would be expressly for Amazon's Prime customers.

It's amazing looking at what's happening in the freight-transportation race. Who would have thought that Amazon could be in both the ocean freight and freight-brokerage space? You'll find out in Chapter 8 that this is happening, and it's been happening since 2016 when Amazon's Chinese subsidiary filed as a freight forwarder. Four years later, Amazon launched its freight-brokerage business. As much as these initiatives employ thousands of people, Amazon is slowly dominating the global logistics landscape.

Tesla and Musk aren't just watching Amazon's domination from the stands. Soon, Tesla will start production of its Tesla Semi, an electrically operated truck. Already, there are a couple of companies that have ordered this unique vehicle. You'll learn more about it in Chapter 8. The Boring Company, one of Musk's companies, will soon start digging tunnels for freight transportation. Discover how this company plans to accomplish that in this chapter.

In Chapter 9, the race between Elon Musk and Jeff Bezos takes us to low orbit. SpaceX's Starlink and Bezos' Project Kuiper are competing to supply satellite Internet to self-driving vehicles. Already, SpaceX is beta-testing its service in the United States, United Kingdom, and Canada. Project Kuiper is still behind, but making progress. Discover details about these two projects in this chapter, and find out what's causing the tiff that's going on between Bezos and Musk regarding satellite Internet.

As you were reading above, a thought probably came to mind: *"Who is this trying to show me how Musk and Bezos are trying to dominate global logistics?"* That's a fair question, and I'm happy to tell you who I am and why I'm writing this book. My name is Dan Bleigh, and I have a Master of Business Administration degree. I've been following Elon Musk and Jeff Bezos for a decade. My admiration and inspiration for these two gentlemen triggered me to turn my passion into written words and write a book. Most importantly, studying these two billionaires over a long time has enabled me to identify their long-term strategies regarding global logistics. It's this I'd like to share in this book so that you're not focused on media hype, but on substance, regarding Musk and Bezos' projects.

This is an exciting journey, filled with revelations of technology and innovation in the logistic space. Most importantly, you should realize why it's critical to think long term. Why not take this journey with me as I analyze and share facts with you throughout the rest of the book?

Chapter 1:

The Size of Global Logistics, and an Introduction to Bezos' and Musk's Companies and Projects Aimed at Dominating Global Logistics

"If you're competitor-focused, you have to wait until there is a competitor doing something. Being customer-focused allows you to be more pioneering." — Jeff Bezos

The global logistics industry is growing due to several factors. A look at the third-party logistics (3PL) market shows it exceeded $1 trillion in 2019 and is expected to grow by at least 9% from 2020 to 2026. Factors driving this growth include a global growth in e-commerce trading and a general increase in globalization. Estimates of the size of the global logistics industry range from $8 trillion to $12 trillion.

Revealed: The History of Logistics and Supply Chain

Logistics and supply chain management have become synonymous with the transport industry. With the advent of technology and growth in consumerism, the traffic industry is growing by leaps and bounds. Where do these two terms come from? The first thing we'll investigate is the origin of the word 'logistics.' There seems to be divergent thoughts on when the word 'logistics' originated. Etymology Online reckons that the word 'logistics' originated in 1846 from the French word 'logistique,' which means "the art of moving, quartering, and supplying troops" (Online Etymology Dictionary, n.d.). It's fascinating that the Romans named the military administrative official 'Logista,' a term that bears close similarities with the word, 'logistics' (Mahnken, 2018).

We have to take a step back to 1838, where we find a book called *The Art of War*, by Baron Henri de Jomini, who was a general in Napoleon's French army. Here, we find the first use of the French word 'logistique' whose meaning is referenced above. However, the first English use of the word 'logistics' appeared in the January 1810 edition of *The Scots Magazine and Edinburgh Literary Miscellany*, referencing Dr. William Muller's book, *The Elements of the Art of War* (Benjabutr, 2020). This seems to hold water in that the 1811 version of the same book called *The Elements of the Science of War* contains the word 'logistics' (Müller, 1811). One thing is clear, the origin of the word 'logistics,' and that's that it comes from military operations. Of course, it later permeated other fields and developed as a branch of industry itself.

The second term of interest is the phrase "supply chain." The *Journal of Commerce Online* attributes the origin of the phrases "supply chain" and "supply chain management" to Keith Oliver, a British consultant. It's thought that Oliver used the terms when interviewed by the *Financial Times* in June 1982 (Prince, 2020). However, authors, such as Mahnken (2018), claim that *The Independent* newspaper used the term "supply chain" in 1905 to describe a war scenario. The same author reckons that the *Glasgow Herald* newspaper used "supply chain" when covering the first Balkan War in October 1912. The phrases "supply chain" and "supply chain management" were first recognized in the late 20th century. Supply chain involves all processes involved to transform raw material into a product and distributing it to various destinations. It appears the origin of the phrase "supply chain," originated from war situations.

Notice that logistics deals with the movement and storage of goods, while supply chain includes sourcing of raw materials, processing them, and delivering the final product. So, logistics can be thought of as a subset of a supply chain. Due to globalization and technological advancements, today's logistics industry is huge.

Amazon: What This U.S. Company Is Really About

In 2015, Bloomberg ran a story that revealed Amazon's ambition to build a global delivery business to rival Alibaba, FedEx, and United Parcel Service (UPS). It's strange that FedEx and UPS partnered with Amazon in delivering its products to its customers. When Brian Olsavsky, chief financial officer since June 2015, was asked about this project, he denied that the Seattle company was intending to go into the transport industry. Bloomberg said that a 2013 report, circulated to senior management, called the project "Dragon Boat." Amazon wanted to be at the forefront of delivering customer goods manufactured in China and India to places like Atlanta, New York, and London.

Although Olsavsky denied claims of Amazon's plans to build a global delivery business, it took a few months to start seeing the truth of the Dragon Boat project. When March 2016 arrived, Amazon had leased 20 Boeing 747 planes from Air Transport Services Group (ATSG), proving the veracity of the 2013 report. Since then, Amazon has gone on to participate in logistics-related initiatives. We'll look at this company in closer detail to understand its activities in the global logistics network.

Over the years, Amazon has extended its tentacles to various industries, including automotive, transport, and health. There are numerous ways that Amazon is involved in the auto and transport industries, such as the following:

- Delivers goods sold from its operations to customers, and uses transport, including air and automotive products, for this purpose.

- Amazon invests in automotive technology to be at the forefront of innovation. Over the years, Amazon has acquired a few startups to strengthen its role as a technology innovator. For example, recently, Amazon announced it has reached an agreement to purchase Zoox, a business that designs ride-hailing vehicles, for $1 billion (Amazon, 2020).

- Amazon Web Services (AWS) is an Amazon product that offers cloud computing services and serves the automotive industry in creating, marketing, producing, and using auto products. A good example of the involvement of AWS in the auto industry is its recent agreement with Ferrari S.p.A. As per the agreement, AWS becomes Ferrari's official cloud, artificial intelligence, and machine-learning provider. The thrust behind the deal is to speed up innovation within Ferrari, including in their Ferrari Formula 1 cars (Amazon, 2021).

The history of Amazon suggests that this Seattle-based company will keep entering new transportation industries. For us to appreciate this potential, we should back up and understand Amazon as a business. The foundation of Amazon is its four guiding principles: long-term thinking, customer obsession, passion for invention, and commitment to excellence. Read any Amazon financial report, you'll always find these four principles in it. Jeff Bezos, the founder of Amazon, preaches these principles daily and has done so since the company's founding. As you read what follows, keep Amazon's guiding principles in mind.

How Big Is Amazon?

Amazon was founded by Jeff Bezos in 1995 as an online retailer. Following his graduation from Princeton University, Bezos worked on Wall Street and other related fields elsewhere up to the time he founded Amazon. Today, Amazon is one of the largest companies in the world. In 2015, Amazon joined a rare list of companies that had made $100 billion in annual sales, and it did so faster than any other company (Gallo, 2021). Amazon has been climbing up the Fortune 500 list of companies. In 2016, it occupied the 18th position, had 230,800 employees, and was valued at $280 billion. Five years later, in 2021, Amazon is well-perched at position two on the list, and, with a revenue growth of 37.6%, this company grew the fastest among the 2021 Fortune 500's list of companies.

In 2020, Amazon's revenue rose 37.6% to $386 billion, including services that brought in $170 billion and product sales that made $216 billion (Juliussen, 2021). (Most of the data in this section is sourced from Juliussen (2021)). Of these sales, the North American market segment contributed 61%, and international sales made up 27%. Nearly 12% of Amazon's sales came from AWS, which was about a 0.4% decrease relative to 2019. However, AWS is by far the most profitable of Amazon's arms with an operating income of $13.5 billion, which is 59% of Amazon's overall operating income.

Exciting for Amazon was that its net income hit $21.33 billion in 2020, a rise of 84% compared to 2019. Amazon's profitability sat at 5.5% of its revenue, a number that's on the low side.

The 2020 pandemic led to an increase in Amazon's e-commerce business, resulting in a 62.7% increase in employees from 2019. Compared to 2016, the number of employees has grown by more than a million, with both full-time and part-time employees. That represents a growth in the number of employees by nearly 465%, or 4.65 times in just five years. If you add in independent and temporary employees, the numbers are staggering.

Amazon hosts third-party sellers on its platform, and these have contributed 20.8% to total sales, which represents $80 billion. Sales from Amazon's online store delivered $197 billion, which equates to 51.1% of net sales in 2020. The United States is the dominating source of Amazon's total sales by contributing 68.3% of its sales, which is bigger than the United Kingdom, Germany, and Japan combined. As such, the U.S. is an important market for Amazon.

Other segments that contributed to Amazon's 2020 total sales include:

- Amazon's subscription services, which contributed $25.2 billion, a 6.5% share of sales.
- Advertising services and other services brought in $21.4 billion, which is 5.6% of net sales.

One of Amazon's 2020 financial negatives is that it spent $61 billion in global shipping costs, a figure that equates to 15.8% of total sales. Shipping costs, as a percentage of sales, have been edging up from 11% in 2015. It's clear that, if Amazon could reduce this cost, it would increase its net-income position, something that's good for its shareholders. That's why the logistics and deliveries industry is so important for Amazon going forward.

Amazon's Logistics and Delivery Strategy

In the beginning of its life, Amazon used delivery services from other corporations. As explained earlier in this chapter, Amazon revised its logistics strategy by adding its own delivery capabilities to the mix. Over the last decade, the Bezos-founded company has been developing its logistics capabilities and is now one of the largest delivery companies in the world. No wonder UPS' share of Amazon shipping has declined from 49% to 22%, while FedEx's dropped from nine percent to less than three percent since 2015. The following summarizes key developments in Amazon's logistics:

What Is Amazon Air?

Amazon Air, previously Prime Air, is Amazon's arm for air transporting customer packages in the United States and certain foreign nations. The business has a fleet of 74 aircraft, and the number is expected to increase with more on order. Initially, Amazon's strategy was to lease aircraft, but has since begun buying its own fleet. In January 2021, Amazon purchased 11 aircraft, which they expect to be running in 2022.

Amazon uses small airports strategically positioned close to package-sorting and fulfillment centers. This helps with optimizing delivery costs. The focus for Amazon is to cover the United States because that's where it makes the most sales.

To the air-logistics strategy, Amazon is adding drone delivery and has named this project Amazon Prime Air. Amazon began its drone project in 2013 with development of drones. In June 2019, the Federal Aviation Administration (FAA) granted Amazon permission to test its drones. For the next stage of the strategy, which is to run autonomous drones, the FAA gave Amazon permission to test package deliveries by using unmanned drones.

For air-safety purposes, Amazon uses sensor-based electronics to prevent crashes from happening. Furthermore, Amazon's upcoming drone-management system will enable low-altitude drone flights by enabling communication and cooperation between drones. It won't matter who's operating the drones; the drone-management system will be designed to do its job. With autonomous drones, Amazon is expected to deliver customer packages faster and closer to its vision of 30-minute deliveries.

Amazon's Ground Transportation Strategy

Amazon uses ground transportation to deliver its e-commerce products cheaper and faster than its online and offline retail competitors. It does this in the following two ways:

- By using long-haul and middle-mile vehicles to transport goods between gateway locations and fulfillment centers and package-sorting locations. The main vehicles used for this purpose are trucks, and, in the future, these trucks could be replaced by autonomous versions.
- Last-mile delivery of goods from Prime Now hubs and fulfillment centers to customer doors. Vans are the vehicles of choice for this purpose,

although there are special electric bikes used in selected cities It's expected that airborne drones will replace vans and bikes in the near future.

The current Amazon's ground transportation strategy is to use third-party logistics networks, largely based on small and independent transport businesses. Amazon says that their Delivery Service Partner doesn't need to have experience because it has the technology, processes, and more than 20 years of logistics experience. This allows truck drivers to own businesses while helping Amazon build a sophisticated third-party logistics network. Truckers can use Amazon Relay, a proprietary truck-brokerage app, to manage transport activities between Amazon and its brokerage-app users.

How Amazon Uses Vehicles

Amazon's middle-mile and long-haul delivery partners move packages in and out of air hubs and to and from fulfillment centers. Drivers can work for themselves or for other companies. Amazon owns nearly 2,000 Prime-branded trucks, mainly from Volvo and Kenworth. The bulk of these trucks feature driver-monitoring systems (DMS) for enhanced safety. Additionally, Amazon has about 30,000 Prime-branded trailers.

Amazon uses numerous fossil-fuel burning vehicles, and negatively contributes a fair share to the environment. To curb this problem, Amazon is quickly incorporating electric vehicles for its fleet of last-mile vans. For example, they ordered 20,000 Mercedes-Benz

Sprinter vans in quarter 3 of 2018. Furthermore, Amazon has begun deploying some of its 100,000 battery-operated electric vehicles it ordered from Rivian. To rival competitors like Walmart, it's likely Amazon will continue to increase its delivery fleet as well as open small fulfillment centers as time goes on.

Other Amazon Projects You Should Know About

One of Amazon's guiding principles is "passion for innovation," and it's living according to this philosophy. Many of the projects that this company is undertaking play some role in the transportation industry. Most of Amazon's innovations create products and services that require lots of chips, electronics, IT systems and even more software—from embedded software clients to cloud software and SaaS.

Alexa is a well-known innovation in homes and many other systems, and it is growing fast in car-infotainment systems. Some things you can do with Alexa installed in your car are: remotely lock/unlock your doors, check fuel level, and start/stop the car engine. Currently, Amazon's Alexa operates in at least 27 car brands, such as Buick, Volvo, Cadillac, and Chevrolet, and 180 of those brands' car models.

Amazon has developed an In-Garage Delivery system that offers a convenient and secure way to receive Amazon's products. The system utilizes one-time codes to open select garage doors for delivery of your packages. Only the delivery person has the code, and, so, only such a person can open your garage for delivery purposes. In-Garage Delivery debuted in April 2019 and served Prime customers in 50 United States cities. By November 2020, Prime members in more than 4,000 cities were utilizing this service. Potentially, there are millions of U.S. Prime members who could benefit hugely from Amazon's In-Garage Delivery.

Not only is Amazon playing a huge part in the auto and e-commerce industries, but it's also extending its innovation into broadband Internet. The conglomerate announced, in April 2019, that it would deploy 3,236 satellites through a project known as Project Kuiper, which will cost around $10 billion. The Federal Communications Commission (FCC) granted Amazon permission to deploy these satellites, and at least half of them will be launched by 2026. Through this project, Amazon hopes to provide plenty of bandwidth for its Prime customers. However, Project Kuiper isn't alone in its mission; SpaceX's Starlink has already deployed more than 1,000 satellites, and plans to send a total of 11,000.

The world is slowly transitioning to a low-carbon economy, and Amazon plans to play its part through its Climate Pledge Fund, with an initial funding of $2 billion. The aim is to invest in companies focused into developing products and solutions geared toward materializing a low-carbon economy. The auto and transport industries will be the major beneficiary of the fund, such as in renewable energy technology, food and agriculture, and transportation and logistics.

Amazon intends to move toward sourcing 100% of energy needs from renewable resources, especially solar and wind. To achieve this major milestone, Amazon will need to supply its operations with more than 18,000 gigawatt hours of power annually. There are already 26 new utility-scale renewable projects in countries, such as Italy, Australia, South Africa, and Germany.

Elon Musk Is Disrupting These Industries

Elon Reeve Musk, a South African-born American entrepreneur, co-founded PayPal and founded SpaceX, an aerospace transportation company. Musk, who left his home country of South Africa in 1988, obtained a bachelor's degree in economics and another in physics. However, he dropped out of graduate school at Stanford University to participate in the 1995 Internet boom. He is one of the early investors in Tesla, an electric-car company, and is now its chief executive officer (CEO). Musk occupies position two, behind Jeff Bezos, on the list of the richest individuals in the world (Blystone, 2021).

Musk thinks big, and, for that reason, his innovations are disrupting the world in many ways. For example, through SpaceX, Musk achieved what was once thought impossible, safely landing a rocket back on Earth without destroying it like many did before him. It's rare to listen to conversations about the future of space and electric vehicles without mention of Musk.

The Automotive Industry

Musk's company, which is taking the car industry by storm, is called Tesla, at which he is a CEO and product design lead. This company, although it missed production targets in the past, has grown to become the biggest carmaker in the world.

Why Was Tesla Founded?

Tesla was founded in 2003 with the aim of transitioning the world toward sustainable energy much quicker. To understand what Tesla is trying to achieve, imagine a self-driving car that you can rent out to someone and never worry about filling it with gas! This is the future that Tesla is envisioning. It has not been a smooth-sailing process to accomplish this ambition. There have been production issues that delayed delivery of vehicles to customers, and this raised concerns among some of Tesla's shareholders.

Partly because of this issue, Tesla has attracted attention from short sellers, who predicted that its shares would keep dropping in value. Unfortunately, these short sellers were beaten to their game and lost $40 billion owing to its stock price rising by 743% during 2020 (Isidore, 2021).

Why Electric Vehicles Are the Way to Go

Electric vehicles are slated to become mass market, judging by what countries are doing to move toward renewable energy. For instance, Great Britain has decided to phase out gas and diesel cars from 2030, which was brought forward by 10 years. The British government intends to inject $3.7 billion to build gigafactories and expand the electric-charging network (Vetter, 2020). China, one of the biggest consumers of electric vehicles, wants one out of five cars in that country to run on alternate fuel (Bradsher, 2017).

As more and more developed countries lead the way toward electric vehicles, the public will gradually adopt these cars. Eventually, any carmaker lagging behind in developing electric cars may go out of business.

Tesla Vehicles' Role in Developing Electric Cars

A typical American pays about $2,100 annually to fill their car with gas and to cover motor oil expenses. Trucking companies spend nearly $200,000 annually to fuel their semi. Replacing your internal-combustion-engine car with an electric car could save you thousands in gas per year. Furthermore, going electric on cars can help reduce the economic burden and reliance on oil.

Although Tesla electric vehicles rely on the grid for charging, they could help lighten demand for oil and gas and decrease the country's carbon footprint. Tesla cars, in addition, are fitted with self-driving technology for autonomous operation. They are fitted with eight external cameras, 12 ultrasonic sensors and radar to model the surrounding environment. Using those models, a Tesla can activate auto-steering, summoning, and self-parking. However, these cars are not yet fully autonomous. One burning issue about Tesla vehicles is that their autonomous driving system can cause crashes.

Tesla's Rough Ride in the Beginning

The Tesla Model 3 is the best-selling electric car in the world and was so in 2019 and 2020. It has a 12% market share, equivalent to the combined market shares of the next four models. In 2020, Tesla sold 365,240 Model 3 vehicles, about 250,000 units higher than second-placed Wuling HongGuang Mini EV. However, getting to this position wasn't an easy ride for Tesla.

Tesla announced it would be producing Model 3 in March 2016 and received nearly 250,000 preorders, valued at $10 billion in sales. Delivering those vehicles proved to be a thorn in Tesla's side. Tesla promised to produce and deliver 1,500 Model 3 units in quarter three of 2017 and ramped up production to 20,000 units per month by December. Tesla delivered only 260 units in the third quarter of 2017, almost 83% below target. Due to unforeseen production problems, Tesla shifted its 20,000-unit-per-month target to March 2018. Still, they could not deliver on their promise, selling 100,000 units that year.

Musk did not give up, and his company nailed its 20,000-unit-per-month target in June 2018. For this to happen, Musk had to do three things:

- He took just two weeks to build a tent to house a completely new assembly line for production of the Model 3 sedan.
- Stayed at the production facility nearly the whole day daily and worked 120-hour weeks to assist with unchoking bottlenecks.

- Sourced employees from other business departments to help speed up production.

Since that time, production of its vehicles has been up, and Tesla delivered 90,650 cars in quarter two of 2020 and went on to produce nearly 500,000 units that year. This California-based company intends to produce one million vehicles in 2021.

Tesla Enjoys More Success

There are a number of developments that helped Tesla ramp up production of its electric vehicles. Tesla's China manufacturing plant nailed its targets of producing 1,000 Model 3 units per week in December 2019. The revenue generated helped this company ride out the impact of the closure of its U.S. plant due to the 2020 COVID-19 pandemic.

In early 2020, Tesla started building its third vehicle-assembly facility in Germany. It was all smiles within Tesla as the electric carmaker delivered its newest Model Y, an SUV, ahead of time. In July the same year, Tesla announced that it had made profits in all four previous quarters, something they hadn't done in their 17-year history. In late 2019, Tesla revealed it would build an electric pickup called Cybertruck. Despite things looking rosy, there are skeptics wondering if the company can keep delivering profits going into the future. One of the concerns is that the expired U.S. tax breaks for buying Tesla's electric cars may weaken demand.

SpaceX, Elon Musk, and the Telecommunications Industry

Elon Musk isn't creating many original ideas; he finds old ideas that failed to work in the past and brings bright engineers and the right technology to make them work. That's what Musk is doing with the Mars project through Starship. He's doing it again in the $1 trillion satellite Internet industry through Starlink

For all the talk of Musk's innovation, his average project seems to revolve around a set formula: find an old idea that failed because of lackluster technology, and attack it with some of the world's best engineers.

That's exactly how Musk and SpaceX are going after the $1 trillion satellite Internet industry to provide online access to those in rural, hard-to-reach locations.

What Is Starlink's Mission?

To understand what Starlink is about, here is a quote from their website: "Starlink is ideally suited for areas of the globe where connectivity has typically been a challenge. Unbounded by traditional ground infrastructure, Starlink can deliver high-speed broadband Internet to locations where access has been unreliable or completely unavailable" (Starlink, n.d.). So, Starlink isn't competing with traditional telecommunication companies, but developing a niche to improve Internet access to people who have traditionally been marginalized due to factors like cost.

SpaceX, Starlink, and the State of Satellite Internet

Satellite Internet is not a new concept; Teledesic ventured into building a constellation of satellites to provide a wide network of broadband Internet. However, it and many similar projects failed to take off due to the cost and logistics of transporting thousands of satellites into space. Furthermore, with little success, some of these companies delivered satellite Internet with high-latency. This means that the quality of videos was low because of slow delivery of data packages from one point to another.

Enter SpaceX and Starlink, and you have a combination of cost-effective transportation of satellites to space and a satellite network company to make satellite Internet viable. Starlink differs from similar past companies in that it's targeting a satellite Internet maximum allowable data transfer rate at one gigabyte per second (Gbps) as opposed to the 25 megabytes per second that others targeted. Furthermore, current high-orbit Internet delivers data at more than 600 milliseconds (ms), way above the 20 ms that Starlink is intending to achieve. They'll do this by orbiting their satellites at a lower altitude than current providers.

Starlink deployed two satellites, Tintin A and B, into orbit in February 2018. By August 2020, Starlink had sent about 600 satellites, 200 less than it needed for the network to become operational. By May 2021, SpaceX had launched nearly 1,800 satellites for broadband Internet. Starlink became operational once it hit its target number of satellites, and, by February 2021, it had 10,000 users from North America and internationally (Reichert, 2021).

It's not enough to send satellites into orbits; you also need to have a de-orbiting plan. Over time, satellites deteriorate and could cause a series of devastating satellite collisions. Already, there is about 6,000 tons of material in low Earth orbit, and SpaceX shouldn't contribute to the issue. As we covered already, Project Kuiper is following in Starlink's footsteps by deploying its own satellites for broadband Internet.

Elon Musk and His Revolution of the Energy Sector

One of Elon Musk's priorities is to eliminate our dependence on fossil fuels and shift to using solar energy. Musk has been chasing after this dream for more than a decade.

The Company Behind Musk's Solar Energy Generation

Elon Musk conceived a company called SolarCity and shared his vision with his cousins, Lyndon and Peter Rive, in 2004. This company was at the forefront of the early 2000s' rush toward solar-energy generation. SolarCity became the largest provider of residential energy before suffering public financial problems. Tesla came to its rescue by purchasing it for $2.6 billion in 2016; however, the purchase has appeared to be a veiled bailout, which has attracted a shareholder lawsuit.

Having said that, SolarCity has become an important element in Tesla's revolution of the auto industry.

True to Musk's daring innovation style, SolarCity wanted to make solar energy mainstream. It took nine years for SolarCity to become the largest installer of residential solar systems in the United States. This company achieved this by revolutionizing the accounting side of solar energy. It cost between $30,000 and $50,000 upfront to install a solar roof. To reduce the cost barrier to purchase a solar system, SolarCity developed a financing system in which homeowners could get a free installation and pay back the cost via installment. This approach worked like a charm, since more than three-quarters of new solar installations from 2014 were solar leases. Unsurprisingly, SolarCity's stock price peaked in February 2014.

However, SolarCity experienced cancellation rates as high as 45%, due to aggressive sales tactics. Some customers decided to exit their lease contracts when they realized that they were not making the savings they were promised. Following a bad start in February 2016, SolarCity's stock price declined by a third of a percent, and that's partly why Tesla acquired this renewable energy provider.

Tesla's Solar-Energy Future

Elon Musk made a bold promise at the 2017 National Governors Association meeting in Rhode Island. He said that the solar technology from SolarCity, working in concert with Tesla Powerwall, could supply enough power for the entire United States from a 100-square-mile piece of land. We're still waiting for that promise to materialize.

Tesla took Solar Roof preorders in 2017 and were first installed to non-employees from the spring of 2018. This is what Tesla promises with its solar-roof technology: "Solar Roof is the only roof that can help pay for itself with the energy you produce. Power your home at the lowest price per watt of any national provider and take control of your monthly electricity bill" (Tesla, n.d.).

Elon Musk Introduces a New Mode of Human Transport

As said earlier, Musk has a knack of taking old ideas and turning them into reality. The idea of transporting people with a vacuum tube first surfaced in 1812 from the brain of an Englishman named George Medhurst. Many people thought that the idea was a pipe dream. However, in 2012, Elon Musk explained the idea so easily that people began to grasp it.

The Hyperloop: The Fifth Mode of Transport

The first reliable mode of transport for humans was their legs, then came horses before cars took over. Over the years, we developed planes, trains, and boats as forms of transport. Many would have thought that no new mode of transport is possible. However, Elon Musk had different thoughts; he introduced the idea of the hyperloop in 2012, then detailed how the technology could work. Musk envisioned people traveling in a low-pressure tube supported by air and propelled by a magnetic linear accelerator.

Bringing the Hyperloop to Life

Musk didn't just dream about the idea, but went on to work on it with SpaceX and Tesla teams to figure out if it could be feasible. From this work, the teams produced a white paper claiming that the hyperloop could complete a 30-mile journey in a mere 2.5 minutes, and thus, reduce a six-hour trip to 30 minutes. Furthermore, the team showed that this trip would cost $20 each way for the system to sustain itself.

Industries to Be Impacted by the Hyperloop

There are numerous industries that could be impacted by the hyperloop, including the airline and real estate industries. Traveling via a hyperloop is faster and cheaper than doing so with an airplane. As a result, people could opt to use the hyperloop instead of airplanes, and thus, negatively impact the more-than-$765-billion airline industry. Furthermore, the speed of the hyperloop could lead to people changing where they live and impacting both commercial and residential real estate.

Another industry likely to be impacted is freight shipping. Nearly half of goods imported into America flow through Los Angeles and Long Beach ports. From here, more than 14,000 trucks deliver those goods to warehouses and rail yards. These trucks burn around 68 million gallons of fuel annually to transport 11,000 containers daily. Trucks would still be needed for last-mile delivery, but the hyperloop could take over transportation of goods from ports to warehouses. This would help decrease air pollution since fewer fuels will be used.

Businesses Working on Hyperloop Systems

Musk may have reignited interest in the hyperloop, but he's not actively involved in making his hyperloop vision a reality. Instead, other companies have taken over the baton and started working on the hyperloop in earnest, including the following:

- TransPod, a Canada-based company that unveiled plans to build a two-mile test track in France in 2019;
- Hyperloop Transportation Technologies (HyperloopTT), which has deals with France, China, and Ukraine to construct hyperloop systems;
- Virgin Hyperloop One, a company that is working on hyperloop projects in the U.S. and India. It is the first hyperloop business to construct a test track and transport two people; and
- Netherlands-based Hardt Hyperloop, whose objective is to operate commercially in 2028.

In a move that's likely to move the hyperloop and railway industry forward, the U.S. Department of Transportation has placed hyperloop proposals under the care of the Federal Railway Administration (FRA).

This department is the first to provide such a guidance, making it possible for future projects to receive funding from the FRA.

How Elon Musk Joined the Tunneling Industry

In December 2016, Elon Musk was stuck in traffic outside Los Angeles (L.A.), and got frustrated with what he was seeing. He tweeted that he was going to build a tunnel-boring machine and use it to dig a tunnel. His idea was to find an alternative solution to frustrating traffic, especially during peak hours. That thought gave birth to The Boring Company (TBC). Its aim is to build a utility and freight tunnels that are safe and provide low-cost transportation.

The major challenge with boring tunnels is that it costs an arm and a leg. To dig a mile, you'll need to spend from $100 million to $1 billion. However, TBC figured two ways to drop this cost down to $10 million per mile. Instead of constructing a 28-foot-wide tunnel, TBC digs 14-foot-wide tunnels, saving millions of dollars. Musk's dream is gradually taking shape in LA. The TBC constructed, commissioned, and put the Las Vegas Convention Center (LVCC) loop into operation in the first half of 2021. Instead of transporting passengers via electric cars on skates, the loop uses Tesla vehicles with drivers behind the wheel.

One of the challenges with tunneling is production of dirt as waste. To solve this problem, TBC will transform dirt from tunnels into affordable bricks, costing around 10 U.S. cents per brick. In an unofficial strength test, a YouTuber named Dan Markham discovered that Musk's brick is stronger than a traditional red brick (Delbert, 2020).

Other Projects The Boring Company Is Undertaking

The TBC has already completed three of the four projects it was undertaking. The first is the LCCV loop explained above, and the Vegas loop is in the pipeline. This loop will include the existing LVCC loop and future extensions to places like McCarran International Airport and downtown Las Vegas. This loop is envisioned to conveniently transport visitors and the Las Vegas community.

The third project the TBC has completed is a research-and-development tunnel in Hawthorne, Calif. It's a 1.14-mile tunnel intended for testing TBC's public transportation systems, hyperloop, and loop. It cost The Boring Company less than $10 million to accomplish this project. The fourth project, also completed, is called the Hyperloop Test Tunnel. It was designed in 2015 and construction was finalized in October 2016. Lessons learned in designing and building the Hyperloop Test Tunnel served to help build better tunnels in the future.

Why Musk Is Interested in Artificial Intelligence (AI)

OpenAI, a company co-founded by Elon Musk, released the third-generation Generative Pre-trained Transformer (GPT-3), a machine-learning model you can use to generate text from Internet data. This model was trained on 175 billion parameters, almost 117 times more than its predecessor. GPT-3 can be used for various tasks, including generating guitar tabs, images, short stories, or computer code. Why would OpenAI develop such a model?

OpenAI: What Is It?

OpenAI began life as a non-profit organization, but turned around and created a capped profit arm in March 2019. Its main reason for the change was to attract funding so that it can realize its mission. "*What's the company's mission,*" you may ask. The answer is this:

> OpenAI's mission is to ensure that Artificial General Intelligence (AGI)—by which we mean highly autonomous systems that outperform humans at most economically valuable work—benefits all of humanity. We will attempt to directly build safe and beneficial AGI, but will also consider our mission fulfilled if our work aids others to achieve this outcome" (OpenAI, n.d.).

OpenAI was co-founded by Elon Musk and four other participants. Musk is known for being vocal about the dangers of AI and donated $10 million to the Future of

Life Institute in 2015 to run a global AI research program. In the same year, he co-founded OpenAI to strengthen research. Elon Musk stepped down from the company's board in 2018 to focus on his other ventures and to avoid a potential conflict of interest with Tesla's AI goals.

Why Is AI a Threat to Humans?

There are two sides to the AI story: One side sees it as a positive step for improving humanity's lives, and the other perceives it as a potential cause for the demise of humans. Elon Musk is among those who think AI threatens human existence. The opposite group, including Google, Facebook, and Amazon, believes AI will lead to less work for humans, higher productivity, and higher efficiency and will result in an improved quality of life.

When you have AGI that's super-intelligent, doing nearly everything for humans, why would you need it on this planet? Imagine a bot whose function is to keep a certain area of the planet clean. There are two ways of keeping that area clean—by always cleaning up the mess humans make or eliminating humans. Developers aren't looking for the second solution (as far as we know), but you can't completely rule this out. Why do people develop nuclear bombs?

OpenAI has developed a charter to help guide all its operations. Chief among these guidelines is the desire for long-term safety. To this effect, OpenAI commits to making AGI safe, and to encourage the AI community to follow suit. Furthermore, OpenAI commits to the following:

> We are concerned about late-stage AGI development becoming a competitive race without time for adequate safety precautions. Therefore, if a value-aligned, safety-conscious project comes close to building AGI before we do, we commit to stop competing with and start assisting this project. We will work out specifics in case-by-case agreements, but a typical triggering condition might be "a better-than-even chance of success in the next two years" (OpenAI, n.d.).

What's Elon Musk Doing in Health Care?

There's no doubt that Elon Musk is a big thinker, judging by his daring programs such as spaceships to Mars, building autonomous cars, digging tunnels to solve surface traffic issues, and developing cheaper satellite Internet. All these initiatives are visible to the naked eye, but he's also taking on one that's microscopic in the health-care industry.

Neuralink: Linking the Human Brain to a Computer for Faster Thinking

We spoke earlier about artificial intelligence and its impact on the human race, especially the possibility of making them irrelevant. Musk is already aware of this and is doing something about it. He wants to see humans keeping up with computer-operated machines.

This itself doesn't fully solve the AI problem, but it could make humans dependent on others for their livelihood. Imagine what could happen if the computer your brain is hooked to fails! So, the computer has to be fail-safe and independent of the skill of another individual; what if they die?

The biggest risk humans face from AI is that we could become like pets. The worst thing that could happen is a complete elimination of mankind. So, Musk thought that his company, Neuralink, could help prevent this from happening. He envisions a system that can augment human intelligence and hooks the human brain to the digital world to stay ahead of AI.

Chapter 2:

The Tip of the Iceberg:

The Space Race

"You get paid in direct proportion to the difficulty of problems you solve." — Elon Musk

After the close of World War II, a new war, called the Cold War, broke out between the United States and the Soviet Union. This war eventually shifted to space, where each nation wanted to demonstrate its technological and military superiority over the other. The Soviet drew first blood when they sent their R-7 intercontinental ballistic missile-launched Sputnik, the first man-made object to be put in Earth's orbit. This development didn't please many Americans, and intelligence-gathering took off in the United States to understand Soviet military activities. The following year, in 1958, the U.S. launched Explorer 1 under the direction of a German-American rocket scientist named Wernher von Braun. In the same year, President Dwight D. Eisenhower launched NASA, and, over time, America became dominant in the space race with the fall of the Soviet Union. Historically, the space race tended to be between nations, until it shifted to private companies in the 21st century. This move was partly aided by the unfortunate death of seven astronauts when, upon their return from the International Space Station, the Space Shuttle Columbia disintegrated when reentering the Earth's atmosphere.

Elon Musk and Jeff Bezos top the list of the world's most powerful chief executive officers. Bezos spent his early life in Houston and Miami, and he holds a degree in electrical engineering and computer science. Both Bezos and Musk have interest in space exploration and travel, and have companies to pursue those interests. The space race is a big deal, financially. To illustrate, the global space economy in 2020 was worth $371 billion, nearly a 1.5% increase compared to 2019, and is expected to grow moving forward.

Elon Musk, SpaceX, and Aerospace

Elon Musk believes humans shouldn't live on earth, only because that could lead to their extinction. To improve our survival, Musk suggests that it's important to explore and, possibly, inhabit other planets. Some of the concerns Musk raises that could lead to human extinction include depletion of natural resources here on Earth.

Enter SpaceX

Elon Musk decided to act on his belief about human extinction and founded SpaceX to make humanity multiplanetary. When you open the SpaceX website, you're greeted by this Musk quote:

"You want to wake up in the morning and think the future is going to be great — and that's what being a space-faring civilization is all about. It's about believing in the future and thinking that the future will be better than the past. And I can't think of anything more exciting than going out there and being among the stars" (SpaceX, n.d.).

SpaceX is the first private company to transport two National Aeronautics and Space Administration (NASA) astronauts to and back from the International Space Station (ISS). This was an important development since NASA stopped its U.S. space shuttle program in 2011.

One of the challenges for expanding space travel is high launch costs. The major drivers of these high costs include capital requirements for building single-use rockets, high system complexity, and low failure tolerance. Reusable rockets are the key to make space travel affordable and accessible. Presently, "expandable launch systems" are the most expensive. Two good examples of these types of systems are the Arianespace's Vega launcher and Boeing/Lockheed Martin Atlas V. Although these rockets can put a lot of cargo into space, they can't be reused. NASA's Space Shuttle, designed to be cheap and reusable, sits in the middle of the cost range. However, expendable solid rocket boosters and the main fuel tank spiked the cost and restricted the value of the Space Shuttle program.

The lowest cost rockets hail from SpaceX, and these are called Falcon rockets. These rockets have already reduced the cost of landing a spacecraft into orbit by up to five times. No wonder SpaceX recently launched its

100th rocket, and has managed to reuse a single orbital rocket six times. Currently, SpaceX has launched more than 120 rockets and reflown 68 of them. Musk's goal is to put one million people on Mars by 2050.

What SpaceX Is Doing to Reach and Return From Mars Safely

Improving the cost of commercial interplanetary space travel will take more than reusable rockets. This is one improvement, to economically take people to Mars and back, and SpaceX has already succeeded with this step. The other steps are to refill rockets in orbit and the ability to manufacture the appropriate fuel on Mars. SpaceX is developing Starship, a reusable transportation system, and combines a spacecraft and super heavy rocket to propel cargo to Mars and the Moon. Starship uses rocket engines propelled with methalox, a fuel mix of methane and oxygen. SpaceX hopes to collect carbon dioxide and water on Mars to produce methalox for its return trip to Earth. When completed, Starship will transport up to 100 people at a time. SpaceX is working on plans to transport Japanese billionaire Yusaku Maezawa and his guests close to the moon in 2023.

Jeff Bezos, Blue Origin, and Aerospace

Jeff Bezos founded Blue Origin, a private spaceflight company, in 2000. Here's how Blue Origin (n.d.) describes why Bezos founded this company:

> Blue Origin was founded by Jeff Bezos with the vision of enabling a future where millions of people are living and working in space. In order to preserve Earth, Blue Origin believes that humanity will need to expand, explore, find new energy and material resources, and move industries that stress Earth into space. Blue Origin is working on this today by developing partially and fully reusable launch vehicles that are safe, low cost, and serve the needs of all civil, commercial, and defense customers.

You can see clearly that both Blue Origin and SpaceX are competitors in the space industry. Bezos and Musk dreamed of launching rockets and landing them safely on Earth, thus creating reusable rockets, and both have achieved this objective. These two successes have led to a decrease in costs for space travel. In the first 10 years of its existence, Blue Origin worked stealthily, and that changed in 2010.

The Musk-Bezos Space Rivalry

There are two space-related spheres in which Blue Origin and SpaceX compete: space tourism and NASA missions, and satellite Internet. Musk's company Starlink takes on Bezos' Blue Origin Project Kuiper to expand the reach of satellite Internet to people who aren't accessing it yet. We'll explore the race between Starlink and Project Kuiper in Chapter 9. In this chapter, let's focus on the rivalry between Musk and Bezos regarding space tourism and NASA missions to the space station.

It's unclear when the rivalry between Jeff Bezos and Elon Musk began. According to Christian Davenport (The Space Barons: Elon Musk, Jeff Bezos, and the Quest to Colonize the Cosmos, 2019), the two met at a dinner some time in 2004. Their conversation centered on rockets and space travel. At that time, Blue Origin was roughly four years old, while SpaceX was in its second year of existence. However, SpaceX was moving faster than Blue Origin. So, during that fateful dinner, Musk apparently tried to advise Bezos, explaining the strategies SpaceX had tried and failed. Musk was trying to guide Blue Origin so that it wouldn't make the same mistakes that SpaceX committed. However, Bezos had none of it, primarily because he wanted his company to try and fail, something he wasn't too worried about as a long-term thinker.

SpaceX Wins NASA Contract to the Moon

In April 2020, the National Aeronautics and Space Administration (NASA), awarded SpaceX a contract to land the next Americans on Moon as part of the Artemis program. The decision followed a shortlist in which NASA had chosen Blue Origin, SpaceX, and Dynetics to submit designs for a human lunar lander. During the Artemis program, history will be made when at least one of the astronauts will be a woman and another a person of color. The journey to the lunar surface will consist of two legs, the first will take four astronauts aboard the Orion spacecraft to lunar orbit. There, two of the astronauts will board the SpaceX human-landing system and spend a week on the Moon's surface.

This contract to SpaceX is worth $2.89 billion. This is also history, since it will be the first time a private company launches NASA astronauts to the moon. SpaceX's human-landing system is based on Starship and is designed to handle the moon's surface. Its design relies on the tested SpaceX raptor engines and heritage of the Falcon and Dragon rockets. With this design, the Starship can be reused for launching to Mars, the Moon, and other locations.

Following the award of the above contract, Jeff Bezos' Blue Origin filed a protest challenging NASA's decision to award the contract to SpaceX. Blue Origin alleges that the awarding of the $2.89 billion to SpaceX was flawed because NASA "moved the goalposts in the last minutes" (Duffy, 2021). The bone of contention is that NASA apparently negotiated a price with SpaceX, but not with Blue Origin. NASA had previously said that it would award the contract to two companies, but reneged on its decision and cited budget constraints and lack of funding from Congress as the main reasons for the turnaround. Bezos and Blue Origin didn't like the reasons NASA gave and decided to file a lawsuit against them; and NASA froze the contract until November 1, 2021.

Blue Origin's First Passenger Spaceflight

Bezos first founded his own rocket maker, Blue Origin, in 2000 to realize his dream of enabling millions of people to live in space at an affordable shuttle ticket price of $200,000 per passenger—less than half of the estimated $500,000 that Musk's ticket cost. Musk founded his space company SpaceX in 2002, and the two-billionaire 'fight' went from an intergalactic battle of rocket sizes (Bezos won exclusive use of NASA's launch pads) to Musk's victory as the first private company to put a person into orbit.

On July 20, 2021, Blue Origin's New Shepard rocket launched its first crew 62 miles above Earth. This was a momentous event for Jeff Bezos and Blue Origin in their quest to launch tourists to space. Jeff Bezos became one of a few people to look at Earth from above. Although New Shepard had launched more than 15 times in the past, it was sending passengers to space for the first time. By doing this, Blue Origin is making headway toward catching up with space-traveling private companies such as SpaceX.

Bezos wasn't alone in this historic flight to space, he was accompanied by his brother Mark; Oliver Daemen, an 18-year-old Dutch high school graduate; and an 82-year-old woman named Wally Funk, who was denied the opportunity to do so in the 1960s because of her gender. The passengers had about three minutes of experiencing weightlessness and watched the curved Earth from above. Bezos had the opportunity to fly two weeks after he stepped down as CEO at Amazon.

Bezos became the second individual to launch, via his private space vehicle, behind Richard Branson, the founder of Virgin Galactic. Branson didn't reach the Kármán line, reaching only about 55 miles above Earth. Blue Origin is building another reusable rocket called New Glenn, named after John Glenn, an American who orbited Earth.

Elon Musk Sends Astronauts to the International Space Station (ISS)

Elon Musk and Jeff Bezos are each on a mission to migrate humans from Earth to some distant planet to ensure their survival. Prior to Bezos' launching in Blue

Origin's New Shepard in 2021, SpaceX, Musk's company had already completed the feat about a year earlier. In May 2020, SpaceX successfully flew its first crewed flight for NASA, carrying astronauts Bob Behnken and Doug Hurley to the ISS as a test flight. Nearly a month later, SpaceX launched its Crew Dragon carrying four astronauts to the ISS and docked successfully on April 24, 2021. At the time of docking, the U.S.-built Harmony module linked up with the Crew Dragon at 264 miles above the Indian Ocean. SpaceX will fly another crew to the ISS on October 23, 2021. However, prior to that, SpaceX is set to launch four space tourists in its Crew Dragon before September 15, 2021, for a three-day orbital mission.

Up to this point, SpaceX is dominating the private space race against Blue Origin. The battle between these two richest persons in the world is not only about space exploration, but also about satellite Internet. Musk has gone so far as to call Bezos a 'copycat' following Bezos' company's inauguration of Project Kuiper. The rivalry of these two giants of business is also shifting to the autonomous vehicle segment.

Chapter 3:

Who Will Dominate the Logistics Race Between Musk and Bezos?

*"Friends congratulate me after a quarterly-earnings announcement and say, 'Good job, great quarter.' And, I'll say, 'Thank you, but that
quarter was baked three years ago.'"* — Jeff Bezos

Amazon's online retail business is growing and requires the use of effective logistics to deliver customer packages faster. One of Amazon's four guiding principles is innovating and simplifying, and this is why this online retailer is making strategic acquisitions and investments. All the acquisitions and investments are within innovative companies that complement Amazon in its business operations. Tesla faced mammoth production and logistics challenges when it was ramping up production of its Model 3 sedan. These problems were good in that they forced this electric carmaker to innovate. In this chapter, you'll learn how Tesla was able to overcome these problems and how it dealt with its logistics issues.

Introduction to Amazon Robotics

In November 2019, Amazon announced that it planned to build a robotics innovation hub in Westborough, Mass. Amazon's plan is to open the hub in 2021, and will hire more than 200 technology and manufacturing personnel. The new hub will be an addition to the current North Reading robotics center, also in Massachusetts. Amazon plans to invest $40 million to develop the Westborough facility, a 350,000-square-foot robotics facility. With this facility, Amazon says it will "design, build, program, and ship our robots, all under the same roof. This expansion will allow us to continue to innovate quickly and improve delivery speed for customers around the world" (Amazon, 2019).

Amazon warehouse workers used to walk 10 to 20 miles a day on concrete floors to ensure customer packages are delivered timeously. Stowers walked up and down long aisles while pushing carts packed with products. When they found space, they stowed the products and scanned them to mark their location in the warehouse. Pickers, walking the same way as stowers, plucked products off those shelves, scanned them, and placed them into carts and pushed them to a desired location for the next steps.

However, that is a thing of the past because Amazon's new warehouses are equipped with robots. These robots carry and move shelves, called pods, to stowers and pickers who are standing on cushions at their workstations. This move by Amazon began in 2012 when Amazon bought a young robotic company called Kiva Technologies and changed its name to Amazon Robotics. It was a momentous occasion for the giant online retailer because this purchase set the foundation for it to build future robots and automate most of its warehouse and sorting activities. Today, Amazon has more than 200,000 mobile robots in its warehouses, working in concert with warehouse employees (Del Rey, 2019). This way, Amazon can fulfill its promises of fast deliveries to its Prime customers. Furthermore, introduction of robots has led to a 40% increase in warehouse inventory (Del Rey, 2019).

Amazon Robotics has designed and put to work robots that work, not only at Amazon's warehouses, but also at its sortation centers. Here, Amazon sorts customer packages by ZIP codes before releasing them for delivery. The sortation process used to be fully manual before robots reduced workers' workload. In October 2018, Amazon deployed more than 800 robots at its Denver sortation center. These robots, called Pegasus after the winged horse in Greek mythology, have simplified the sorting work. Pegasus does most of the footwork required in the sortation centers.

Sortation center workers scan customer packages before placing them on a conveyor belt on top of the Pegasus robot. Using machine-learning algorithms, these robots drive to openings of the correct chutes and drop the package on the slide. According to Amazon, this process takes two minutes to complete.

What You Should Know About Amazon Logistics

In 2015, Amazon launched its logistics business to reduce delivery times to consumers who expect their orders to be fulfilled quickly. Amazon Logistics is a shipping and delivery service designed to complement existing delivery providers, such as UPS, USPS and FedEx. They offer seven-day and same-day delivery options, and use various logistics partners across the country to achieve this, including motorcyclists, local walkers, and cyclists. The logistics business is designed to give Amazon and its sellers repeat business, and as such, it can also create business opportunities for entrepreneurs. Customers value timely and accurate service, and Amazon sellers have an opportunity to continually meet customer delivery needs when they use Amazon Logistics.

Five years after launching Amazon Logistics, Amazon opened a 175,000-square-foot delivery station in the Detroit metro area. The aim was to streamline delivery service in the last-mile and complement capacity and flexibility with delivery capacity. Amazon's delivery stations provide not only faster deliveries for its customers, but also present business opportunities for interested car owners with the Amazon Flex program. A car owner interested in making extra money can download the Amazon Flex app, set up an account, and look for convenient delivery opportunities. Once located, they can go ahead and reserve a block, and if successful, they can deliver Amazon customer packages.

If you have a fleet and want to deliver for Amazon Logistics, you must prove to Amazon that you own at least five freight cars, each with about 200 cubic feet of cargo space. As with any last-mile business, Amazon requires you, as a company, to purchase four types of insurance: $1 million for general commercial liability, $1 million for corporate and automotive liability, $1 million for employee liability, and $25,000 for statutory liability for freight. As far as drivers are concerned, you should ensure that they are equipped with safety policies and procedures and have undergone training. Furthermore, you should have all the documentation to prove that your drivers have received the necessary training. Once you have a company that meets the above-mentioned requirements and signs up successfully, you'll pick up packages at a nearby Amazon facility, use its navigation technology to deliver the packages, and finally, get paid.

Why Amazon Went Into Product Shipping

Many retailers were struggling during the 2020 pandemic, but Amazon met its shipping needs with minimal outside help. This was possible because of the more than $60 billion that Amazon invested in its logistics network since around 2013. Amazon started this journey of self-sufficiency when bad weather and ecommerce growth in 2013 resulted in many consumers not receiving their Christmas packages in time. Over time, Amazon is becoming a truly vertically integrated online retailer. In 2019, Amazon delivered nearly six of every 10 parcels to U.S. consumers from its warehouses, and this represents about 22% of all online retailer parcel deliveries in the United States; this way, Amazon became the fourth-largest shipping provider in the United States (Davis, 2020).

Besides shipping packages to its customers, Amazon also delivers packages for its seller network, commonly called Fulfillment by Amazon (FBA) sellers. FBA is Amazon's storage and shipping service it offers to businesses. What these businesses do is send their products to Amazon's warehouses, and Amazon takes over from there until products reach FBA sellers' customers. However, an entrepreneur should first find products and ship them to Amazon FBA for Amazon to do the rest. Typically, many entrepreneurs order their products from China. It's important to ask your suppliers to put the FBA label on your items so that they know that you are an FBA seller. Many Chinese suppliers are only at the beginning and may think you're selling to someone else.

Businesses can choose to ship their customer packages via Amazon Premium Shipping. Customers receive their parcels within a day or two of ordering. This option is available only to sellers who meet high delivery standards and can help a business score high on delivery time and delivery expectations. The seller must have delivered at least 10 pieces within the last 30 days. The seller must make the delivery on time and maintain a delivery value of at least 97% during this period. A seller must use tracking for at least 98% of shipments during this period. During the normal time slot, most premium members around the world receive their product within one day.

Blue Origin and Airspace Logistics

Blue Origin is making progress with the production of its reusable rockets. The New Shepard rocket recently succeeded with its 62-mile journey above the Earth's surface. Another Blue Origin's rocket is called New Glenn, capable of lifting 90,000 pounds into low-Earth orbit (LEO). This rocket is designed to complete at least 25 flights. There are six kinds of engines that Blue Origin has developed, which are called BE-1, BE-2, BE-3PM, BE-3U, BE-4, and BE-7. The smallest engine delivers 2,000 pounds of force of thrust, while the largest develops 550,000 pounds of force of thrust (Blue Origin, n.d.).

The Birth of Tesla Logistics

Tesla started life as a carmaker catering to the luxury market of electric vehicles, and it became good at it. However, things changed when Tesla began producing its mass-market car, the Tesla Model 3. Initially, Musk promised to produce and deliver 1,630 cars in the third quarter of 2017, something Tesla ultimately failed to accomplish. During that quarter, Tesla managed to produce 260 Model 3 cars and deliver 220. Tesla was producing these Model 3 vehicles for the more than 400,000 orders it has received following the launching of this car. It turns out there were two major problems: battery assembly issues at its Nevada Gigafactory and software hiccups for running an automated parts conveyance system at its car manufacturing plant.

Model 3 production issues continued into the first quarter of 2018, but there was an improvement. Tesla had set a target to hit 2,500 Model 3 cars during that quarter, but delivered 2,000 cars, which was 80% of its target. This can be considered an improvement to the previous quarter, but was still far off the 5,000-per-week target for which Musk and his Tesla team were aiming. The improvement came about because Tesla decided to take over production of battery modules at two sections that were run by a third party. Furthermore, Tesla sought help from Germany so that it could automate its Gigafactory's battery production operations. According to Sage & Taylor (2018), Tesla flew in six planes packed with robots and equipment from Europe to Nevada to speed up its Gigafactory's battery production. As for the conveyance system, Tesla erected a tent outside its Fremont car assembly plant to help with ramping up production. Tesla eventually turned the corner on the production of Model 3 in June 2018 when it hit its target of 5,000 cars per week. The performance continued into the third quarter, as Tesla manufactured 5,300 Model 3 vehicles in the last week of the quarter (Shaban, 2018).

As car production ramped up, Tesla's problems shifted to delivery and logistics of vehicles. Tesla CEO Elon Musk described the situation as going from production hell to logistics hell. To solve the logistics issue, Musk decided that Tesla should build its own transporter and trailer. This decision came about because there was a shortage of trucks and trailers when Tesla desperately needed them. There were two other initiatives Tesla took to meet Model 3 sales demand and deliveries. According to Lambert (2019), Tesla wrote an SEC filing about increasing its logistics capacity. A portion of that filing said:

> As part of Tesla's ongoing logistics strategy to increase its vehicle transport capacity, reduce vehicle transportation time, and improve the timeliness of scheduled deliveries, Tesla agreed to issue shares of Tesla's common stock in connection with its acquisition of certain car-hauling trucks and trailers from Central Valley Auto Transport, Inc. ("Central Valley" or the "selling stockholder"), an automotive transport provider. We are registering these Tesla shares pursuant to registration rights granted to the selling stockholder in connection with the acquisition.

The second initiative was to introduce a car-delivery service called (Tesla Direct). This idea helped Tesla avoid overcrowding at its delivery centers and got cars to customers quicker. Tesla Direct involves a Tesla advisor delivering your car at your preferred location, which could be your work, home, or another convenient place. When the car arrives, the customer signs the paperwork and gets behind the wheel within minutes. It's not surprising that Tesla introduced a service like this; Musk had wanted Tesla's logistics to be as important and impactful as Amazon's.

Tesla Factories You Should Know About

Tesla has four fully operational manufacturing facilities, of which three are in the United States and one is in China.

- **Fremont factory**: This factory is the heart of Tesla's electric-car production, and is located outside of San Francisco. Tesla's more than 10,000 employees work at this factory that occupies 5.3 million square feet of space. Tesla's Model 3, Model Y, Model S, and Model X are all produced at the Fremont factory, as are many electric-vehicle components. Furthermore, numerous Tesla parts suppliers have built facilities nearby.

- **Gigafactory 1**: This production facility is based out of Sparks, Nev. It produces lithium-ion batteries and electric motors for the Model 3 sedan. Moreover, Gigafactory 1 has inherited the production of Tesla's energy storage products Powerwall and Powerpack. Currently, this factory is only 30% complete.

- **Giga New York or Gigafactory 2**: Among its products, Tesla produces solar-energy panels and related components. It does this at its 1.2-million-square-foot Giga New York factory, based in Buffalo. Additionally, Tesla produces electrical components for energy-storage products and Supercharger.

- **Giga Shanghai**: This factory is the third to operate in Tesla's lineup of manufacturing plants. It's the second factory where Tesla assembles both the Model 3 and Model Y vehicles. Giga Shanghai, a 9.3-million-square-foot facility, started operating in 2019 after finishing construction in one year. China is an important market for Tesla's electric vehicles and can be an important distribution center to the whole of Asia and even some parts of Europe.

Tesla is also building two other Gigafactories: one in Berlin, Germany, and the second in Austin, Texas. Giga Berlin allows Tesla to tap into the engineering skills and expertise of the Germans, as well as to supply Europe with its cars. Construction of Giga Berlin began in 2019, and it will become the most advanced high-volume electric-car-production facility worldwide. Initially, this factory will be used primarily for production of Tesla's Model Y. Giga Texas will be used mainly for production of the Cybertruck and Tesla Semi. Tesla chose Austin so that it can distribute Model X and Model Y cheaper to eastern locations of the United States.

Production and Logistics of SpaceX Rockets

SpaceX is a private spaceflight company that designs, manufactures, and launches rockets and spacecraft for governments, military, and private missions. Furthermore, SpaceX provides telecommunications through its Starlink satellite Internet operations, as well as a rideshare service for putting satellites into orbit. Since 2008, SpaceX has designed and produced a couple of rockets, but now focuses on making reusable rockets to lower the price for space exploration.

There are two types of equipment that SpaceX produces for space exploration—reusable rockets and spacecraft. Let's explore four of these machines.

- **Falcon 9**: The Falcon 9 is the world's first reusable rocket that can carry between 8,860 and 50,265 pounds, depending on its destination, whether low-Earth orbit or the International Space Station. This rocket uses Merlin engines that develop 190,000 pounds of force of thrust for propulsion into the sky. It has made more than 120 total launches, 85 landings, and has been reflown more than 65 times (SpaceX, n.d.).

- **Falcon Heavy**: The Falcon Heavy is more powerful than the Falcon 9. It uses 27 merlin engines to develop thrust that's equivalent to that developed by 18 747 aircraft. SpaceX designed the Falcon Heavy to lift about 141,000 pounds of payload to orbit, making it the most powerful operational rocket worldwide. This rocket has made seven total landings and was reflown four times (SpaceX, n.d.).

- **Starship**: Starship is a name that refers to a combination of a super heavy rocket and SpaceX's spacecraft. It is capable of transporting payloads, satellites, and cargo to various space destinations, including the Moon and Mars. Starship's payload to low-Earth orbit is more than 220,000 pounds. For thrust, it uses raptor engines that deliver 440,000 pounds of force. On May 5, 2021, Starship serial number 15 (SN15) completed a successful test flight when it launched to 6 miles above Earth and returned safely.

- **Dragon**: This equipment is a spacecraft that has 16 engines and is used to carry cargo or humans to space. It can carry a launch payload of 13,228 pounds while it is capable of returning 6,614 pounds of cargo to Earth. The Dragon has so far successfully completed 25 visits to the International Space Station and was reflown for 11 missions.

Chapter 4:

Musk and Bezos in the Race to Dominate Artificial Intelligence (AI)

> *"When something is important enough, you do it even if the odds are not in your favor."* — Elon Musk

Artificial Intelligence (AI) is changing the world and rapidly, at that. Companies such as Amazon and Tesla have seen the potential of AI for several years and have been building AI systems to build thriving businesses. Tesla uses AI to develop fully self-driving vehicles, and Amazon is applying the technology to sell more of its wares both online and offline and to move goods quickly. For example, Amazon uses AI to make product recommendations to customers, especially on Amazon.com. Once a customer looks at a book, another product, Amazon's AI, learns about these actions. Next time the customer launches the Amazon website, they get product recommendations. Although Tesla is focusing on developing AI to run autonomous vehicles, it can use this system in many other applications.

Musk and Artificial Intelligence in His Companies

What Tesla Artificial Intelligence Is About

Tesla has been working for years to create automated cars, automated factories, machinery, and construction equipment. The focus of Tesla's AI is not on the hardware, but on AI-powered systems. The major problem with developing fully self-driving cars is the ability to interpret real-world surroundings. It's necessary, therefore, to develop an AI solution capable of directing such self-driving cars to navigate various road obstacles, just as a human does, but with improved efficiency. To achieve this objective, a lot of data is required in order to feed deep-learning algorithms.

Tesla cars that are already on the road collect vast amounts of data that needs processing in order to develop real-world AI. With more than a million cars on the road, Tesla treats each vehicle as a sensor for data collection. Tesla then takes this data, analyzes it, and uses it to improve its machine-learning and AI algorithms, create new algorithms, and send them to the vehicle via the cloud. So, Tesla owners don't drive cars to commute to work or run errands. Instead, they are training their Tesla AI engine while on the go. This strategy puts Tesla ahead of many autonomous vehicle carmakers and will become one of the most effective crowdsourcing training initiatives for machine learning and AI.

During Tesla's Quarter 1 2021 earnings call, someone asked a question about Dojo, a supercomputer that Tesla is building. In response, Musk said that many people see Tesla as an automaker or an energy business. According to Musk, this view may be accurate in the short term, however, in the long term, Tesla should be seen as an AI company as much as an energy or carmaker (Seeking Alpha, 2021). Tesla focuses on AI in two areas: autonomous driving and electric propulsion. One of the important hardware components for this to occur is by having an integrated AI chip. Tesla dropped its partner, NVIDIA, to optimize this chip and has since dropped that company in favor of doing the work itself.

The current AI technology used in Tesla cars is based on unsupervised machine learning. For better performance, Tesla's automotive systems will have two AI chips that perform separate traffic-assessment tasks. These chips are optimized to run at 2 GHz and process 3.6 trillion operations per second. The big idea behind having two AI chips is to evaluate traffic hazards and other situations around the car. The two chips are added for redundancy in making Tesla cars safe. Furthermore, the power supply of the chips and the data feed are also redundant. The cameras of the cars have two separate power supplies to prevent a failure.

Unlike many automakers, Tesla adopts a comprehensive approach to building cars. That's why Tesla builds its own software, neural networks, chips, and hardware. One of the powerful hardware products

Tesla is building is a neural-net computer. The carmaker did this because it couldn't find a suitable one anywhere in the world. Tesla did the same with software to have the necessary tools to build arguably the most advanced real-world AI worldwide. For continuous training of the AI system, it's necessary to have a powerful enough computer to handle the vast video data that Tesla collects. Hence, Tesla developed Dojo, a supercomputer for neural-net training.

Tesla unveiled its in-house chip during its live-streamed AI Day, and it's a clear sign of how serious Tesla is with vertical integration. Tesla's senior director of Autopilot hardware, Ganesh Venkataramanan, passionately explained the power of the chip, called D1. This chip has 362 teraflops of processing power and contains 7 nanometer technology (Bellan & Alamalhodaei, 2021). Tesla's neural-net computer will depend on this chip to process vast amounts of camera-imaging data. This milestone should help Tesla overcome the shortage of chips while simultaneously increasing bandwidth and narrowing latencies.

Elon Musk unveiled a humanoid robot, called the Tesla Bot, that operates on Tesla's AI technology. This revelation is in line with Tesla not just being an electric carmaker or energy company, but also an AI and robotics company. Several people were not expecting Tesla to make an announcement like this. The Tesla Bot is anticipated to carry out mundane tasks such as shopping for groceries.

What Is Neuralink Artificial Intelligence?

Elon Musk is a man with many business interests, from electric cars to solar energy, from artificial intelligence to robotics, from sending people and cargo to space to dominating freight logistics. He seems to not be content with what he's currently doing and has carved a new business for himself called Neuralink, introduced in Chapter 1. Not only is this startup intent on implanting a chip in a person's brain, but also wants to level up human intelligence with super-intelligent computers. Neuralink believes there are many applications for the technology it is building.

Currently, Neuralink is focused on designing and building a chip that can help people with various brain disorders. Elon Musk made it clear that the goal for Neuralink is to solve spine and brain problems, and some brain disorders that come to mind include anxiety, depression, memory loss, insomnia, seizures, hearing loss, and brain damage. The present medical device that's been tested on animals is called the Link. It's placed in the brain cortex, a few millimeters in, but as Neuralink learns more about brain communication, it'll go deeper into other brain areas.

The Link connects with thousands of brain neurons using its 1,024 electrodes. In this way, it'll record brain activity, process the signals in real time, and give feedback. Neuralink wants to start off by helping people with spinal-cord injuries so that they're able to control computers and mobile devices by using their brains. For example, as a user thinks of moving their arm, the Link will record their brain activity, decode the information, and send it to that user's computer via Bluetooth. As Neuralink's AI algorithms adapt and users get more practice, users will be able to control multiple devices.

Robots are becoming increasingly as intelligent as humans. Musk believes that AI will outwit the human species at some future point and that connecting the human brain with computers could be the solution. Perhaps the biggest challenge for Musk and his Neuralink team is sharpening human learning and improving their intelligence. If he succeeds, he will succeed by telepathically exchanging information between human brains and the results of AI algorithms.

Ultimately, Musk aims for an AI that connects the human brain to artificial intelligence by implanting a chip in a person's brain. The goal is to use Neuralink's chip to create a human consciousness that would be linked to artificial intelligence. This will equip people with software and algorithmic methods to simulate the abilities of the human brain, including emotional intelligence. If this happens, the chip will be implanted in the brain to create a kind of symbiosis between humans and artificial intelligence that Musk says can sustain and improve our brains and also help the human species to survive.

OpenAI and its Role in Artificial Intelligence

OpenAI, as explained in Chapter 1, is a not-for-profit organization with the aim of advancing digital intelligence to humanity. The organization hopes its research will focus on human impact rather than profiting individuals and organizations. OpenAI's researchers publish papers, blog posts, patents, and code to share with the world. Furthermore, OpenAI encourages collaboration at the research and implementation level. Here's how OpenAI describes its mission:

> As a non-profit, our aim is to build value for everyone rather than shareholders. Researchers will be strongly encouraged to publish their work, whether as papers, blog posts, or code, and our patents (if any) will be shared with the world. We'll freely collaborate with others across many institutions and expect to work with companies to research and deploy new technologies. (OpenAI, 2015)

According to the OpenAI website, there are numerous co-founders of this organization, including Trevor Blackwell, Vicki Cheung, Andrej Karpathy, Durk Kingma, John Schulman, Pamela Vagata, and Wojciech Zaremba. Elon Musk and Sam Altman served as co-chairs at OpenAI's founding. Financial support for the organization was $1 billion, and it all came from YC Research, Amazon Web Services (AWS), several initial financial backers (Reid Hoffman, Peter Thiel, Infosys, Jessica Livingston, Greg Brockman, and Sam Altman), and Elon Musk (OpenAI, n.d.).

Both Bezos and Musk are represented in OpenAI, and each is likely to benefit from the organization's work.

Recently, Microsoft invested $1 billion in OpenAI and is the preferred commercial provider of cloud-computing services for the non-profit. The two companies agreed to experiment and develop Artificial General Intelligence (AGI) with built-in ethics and trust. OpenAI and Microsoft will develop a computational platform in Azure and will be able to train and run AI models. Over time, this platform will scale to AGI. OpenAI intends on licensing some of its pre-AGI technologies to raise funds to access the computational power it needs for its work.

In September 2020, OpenAI licensed its GPT-3 technology to Microsoft. GPT-3 is a powerful general-purpose language within an application programming interface (API) that OpenAI released in June 2020. Users can use this API to access new AI models that OpenAI builds. Most importantly, the deal has no impact on access to OpenAI's API. Microsoft will be able to use GPT-3 for its services and products. By March 2021, more than 300 applications were using GPT-3 for various purposes, including test completion and powered search. At the time, GPT-3 was generating around 4.5 billion words per day on average. GPT-3 is easy to use; for example, if you give it a text prompt, it can complete the text in everyday language.

Another powerful neural network from OpenAI, following GPT-3, is called DALL-E, a multimodal AI model. Unlike GPT-3 and face-recognition models, DALL-E transforms text into images and does so accurately in numerous instances. This AI model is capable of anthropomorphizing certain objects and animals. I personally tested it to see what would happen. So, when I 'fed' DALL-E with the text, "A lamp in the shape of a pig. A lamp imitating a pig," it returned 30 different images of a lamp in the form of a pig. Although the new model of OpenAI is not a real step toward Artificial General Intelligence, and DALL-E is a new tool with extraordinary new capabilities, the fact remains that there is certainly progress toward AGI.

Bezos and Amazon's Artificial Intelligence

The secret of growing a large business is to increase the number of customers, charge them more, and get them to buy more often. Although it sounds so simple, it could be a challenge to make it happen. Customers return to a business if they feel that their previous visit was worth it. This is where customer service comes into the picture. With customer service, a business can accelerate word-of-mouth marketing and, thus, increase the number of people buying from it. Not only do returning customers market your business for free, but they tend to buy more of your goods and services. This is why Amazon is obsessed with customer service and made it one of their guiding principles. The way to serve your customers better than anyone is to always give them what they want. And, this is where machine learning and artificial intelligence (AI) at Amazon is such a vital cog in its business operations.

The Root of Amazon's Machine Learning and AI

Machine learning and AI today complete tasks that were once thought to be uniquely human. What we're witnessing, currently, is just the beginning because more is still to come. Jeff Bezos and many other people and companies are working hard to ensure people are relieved of mundane tasks. Amazon has worked tirelessly to develop its machine-learning and AI systems and has sold some of these services to such organizations as NASA and the National Football League (NFL) through AWS. This was made possible by Bezos' commitment to AI as a company that has begun to invest in machine learning and artificial intelligence.

These products use machine-learning algorithms to perform tasks, such as flying drones, product detection, and command interpretation, all as well as or better than humans. Major technology companies hoping to harness and market artificial intelligence, including Google, Microsoft, and Amazon, are working on their own deep-learning chips. This work has been very visible in recent years, from the Amazon Go convenience store, which uses machine vision to eliminate checkout queues, to Alexa, Amazon's cloud-based AI assistant. First, we need to understand that not all companies have the same reach and capital as Amazon and that, as technology becomes more accessible, more and more companies will find inspiration in how Amazon uses artificial intelligence (AI).

Machine learning is a subset of AI used to mine data patterns for making predictions. For example, if you buy a product on Amazon.com, such as a book on machine learning, Amazon is able to predict what other similar products you may like, and it makes recommendations for you based on your past purchases. Machine learning took off at Amazon in 1999 when Jeff Wilke, former CEO of Amazon Worldwide Consumer, joined this giant online retailer. Wilke built a team of scientists to study Amazon's internal processes to improve efficiency. It didn't take long for the team to find patterns, and, soon, the path led to machine-learning algorithms.

As Amazon grew, largely due to the growth of Jeff Bezos' ambitions, the need for automated insights grew. Instead of making noise about its improved efficiencies due to machine learning, Amazon kept implementing its newfound way of growing its business operations. One area critical to Amazon's operations and its success are operations in its fulfillment centers. Here, there are thousands of six-foot-tall cuboid shelving units called pods Robots slide beneath each and drag them around for specific purposes.

To an onlooker, the movements of the robots may seem chaotic, but that's further from the truth.

Algorithms direct all the robot movements in the warehouses. Automation is a key driver of Amazon's operations. Lest someone think the robots would replace humans, it'd be necessary to know that, at the same time, Amazon had said that it will hire 300,000 workers, 50% more than the number of robots. That gives you an idea about the size of Amazon and how much it intends to grow.

Amazon warehouse employees pack items into pods to be stored at a dedicated spot in the warehouse. Also, they pick items out of the pods delivered to them by robots. Every time they pick and pack an item, these workers scan the product and the applicable shelf with a barcode for identification by the algorithm. For optimum warehouse operations, the time that workers wait for a robot is vital; the shorter the wait is, the quicker deliveries can be made to customers.

Amazon created Amazon Web Services (AWS) to serve other businesses with their cloud-computing needs. Through AI initiatives, Amazon aims to provide AI to enterprises and developers that may not be able to afford to develop their own systems or employ a capable workforce to do so. AWS' offerings include API services that demonstrate the company's commitment to AI and the role of AI in the future of cloud computing.

One of the important leverages AWS can have is its ability to predict computation demand. Failure to forecast the demand for computation could result in errors and loss of sales. Through monitoring the amount of traffic that each of its customers gets, AWS feeds data to machine-learning algorithms that, in turn, predict where and when it'll need more computation. To improve the efficiency and costs of doing this task, AWS has produced a chip called Inferentia. This silicon chip was designed to speed up deep-learning workloads, helping to move AWS toward its vision of making AI accessible to everyday developers. With Inferentia, developers can improve performance inference in the cloud, simplify integration of machine learning to their business operations, and reduce the cost of inference.

If you were to ask a random person on the street, "What does Amazon do," they would probably tell you that Amazon sells books, phone covers, and computers online. In a way, that person would be right, but only partially. Amazon is largely an e-commerce business, but has dabbled in brick-and-mortar stores, such as bookstores, in the past. This online retail giant has again opened a brick-and-mortar business called Amazon Go.

Amazon Go is a cashierless venture that offers convenience foods, such as breakfast and lunch, in a brick-and-mortar store. This is the third vital area where Amazon is using machine learning and AI. The store should allow you to enter, collect the goods you want, and leave without interacting with anyone. Amazon Go is the latest development into its foray in the food business, following its Amazon Fresh service and acquiring Whole Foods Market chain in 2017. Currently, there are more than 25 Amazon Go stores in U.S. cities, like New York, Seattle, and San Francisco.

Amazon Go stores utilize a proprietary technology known as Just Walk Out. To use these stores, you need to download and install the Amazon Go app on your smartphone. Before you enter the store, you should sign in to the app. As you enter the store, cameras and sensors notice and identify you. As you move around the shop and pick items, sensors and cameras installed throughout the shop notice your activities. If you pick an item from a shelf, the app adds it to your virtual cart, and, if you return that product to the shelf, the app removes it from your cart. When you're done shopping, you scan your app at the turnstile and leave the store. The cost of your items will be billed to your account.

Jeff Bezos encourages Amazon employees to share data across various business functions. This means that Amazon Go started operations having information about its target market, especially if you've ever bought anything on Amazon.com. Amazon collects stacks of information about its customers from Echo devices to smartphone apps, and knows a lot about your buying and viewing behaviors. This is what makes Amazon AI activities powerful and helps the company amass millions of dollars.

When you study Amazon closely, you'll realize that it creates a technology, optimizes it, and, then, offers it as a service to other businesses and organizations. A good example is Amazon Web Services, a profit center that brings a chunk of net income to Amazon. Nothing prevents Amazon from employing the same business model with Amazon Go. In this case, Amazon could offer other retail stores the Just Walk Out technology while it gathers data for its own use. This could be the start of a totally new way of shopping, across the United States and other countries.

Chapter 5:

Will Bezos Beat Musk in the Self-Driving Vehicles Race?

> *"If you can make a decision with analysis, you should do so. But it turns out in life that your most important decisions are always made with instinct and intuition, taste, heart."* — Jeff Bezos

Autonomous vehicles, or fully self-driving cars, started as a dream, and companies such as Tesla and Amazon are in full swing to make them a reality. Like with most technologies, developing full self-driving cars is a process and can take years to refine until it can be safely done. Experts estimate that a quarter of all cars sold in 2035 will be autonomous vehicles, with 40% of them being full self-driving cars (GreyB Services, 2021). This is possible because of the availability of advanced modern-day sensors, machine learning, and artificial intelligence currently being used in cars produced by Tesla and Waymo.

Why Tesla Could Win the Self-Driving Vehicles Race

Transparency group, Plainsite, released a record of emails between California's Department of Motor Vehicles (DMV) regarding Tesla's claims about its Full Self-Driving (FSD) technology. The California DMV makes a point that Tesla's Autopilot is at SAE level 2 automation, which means that it's not fully autonomous. Furthermore, this DMV said that Tesla's FSD is also at SAE level 2.

In a letter to the California DMV on December 28, 2020, Tesla admitted that its Autopilot and FSD were not fully autonomous. Here's what Tesla's representative wrote in his letter to the DMV:

> As you know, Autopilot is an optional suite of driver-assistance features that are representative of SAE Level 2 automation (SAE L2). Features that comprise Autopilot are Traffic-Aware Cruise Control and Autosteer. Full Self-Driving (FSD) Capability is an additional optional suite of features that builds from Autopilot and is also representative of SAE L2. Features that comprise FSD Capability are Navigate on Autopilot, Auto Lane Change, Autopark, Summon, Smart Summon, Traffic and Stop Sign Control, and, upcoming, Autosteer on City Streets. While we designed these features to become more capable over time through over-the-air software updates, currently neither

> Autopilot nor FSD Capability is an autonomous system, and currently no component, whether singularly or collectively, is autonomous or makes our vehicles autonomous. This includes the limited pilot release of City Streets. (Plainsite, 2021)

The big takeaway from the above statement is that FSD is at SAE level 2 and that over-the-air updates will, over time, make it autonomous. Customers buying Tesla cars currently opt for an FSD that gives them additional driver-assist features over and above those provided by Autopilot.

One Model X car owner has raised concerns online about their frustrations with an earlier version of the Tesla FSD. They believe the FSD could work in the stop-and-go traffic of L.A. because that's where it's programmed to operate. However, it isn't user-friendly on country roads, such as those in New England. Concerns like these are probably what led Tesla to work on its latest FSD. The good news is that, by constant and never-ending improvement approach, you can achieve amazing results. I foresee Tesla nailing its dream of having an autonomous vehicle soon, though I'm not exactly sure when.

On September 10, 2021, Tesla began shipping its FSD beta version 10 Thus, the thousands of Tesla car owners who bought the FSD option have started accessing the feature. The V10 follows after four weeks of releasing FSD 9, which Elon Musk had been promising since 2018. In 2019, he had proclaimed that FSD version nine would be available in "a year from now," meaning 2020. That didn't happen. Initially,

Tesla offered the FSD at a hefty price of $10,000; however, it turned the corner in 2021 and made it a subscription-based offering. This subscription offer is available for owners of Tesla cars running FSD computer 3.0 or newer versions of Autopilot. This means that car owners who bought their Tesla vehicles in 2019, following the introduction of the FSD computer 3.0, can have access. However, owners with the FSD computer 2.0 or 2.5 may choose to upgrade and access the subscription pricing model.

The latest FSD version is not yet ready for public rollout. It differs from the previous version in that it addresses most known issues, especially with decision-making. Furthermore, FSD beta 10 features improved self-driving graphics over the previous version. However, it's possible that new issues might arise, hence, the reason why it's a beta version. The key, according to Tesla, is keeping safety at the top of your mind when you use the latest FSD version. You never know when the system could do the worst thing at the wrong time.

No one should expect a beta version to be perfect, but the more it's used, the better it becomes. Developing a fully self-driving system is like traveling to the space station. On paper, the path may look straight, but, in reality, a spaceship course corrects many times along the way until it reaches its destination. Developing beta versions FSDs, followed by testing, is probably the quickest way to get to fully autonomous vehicles. Tesla and the team behind the journey to autonomous vehicles should be applauded for willing to put their

FSD on the roads. However, it'd be helpful if Musk could deliver what he promises on time.

How Tesla Plans to Solve the Shortage of Trucks in the United States

The shortage of truck drivers is weighing heavily on road freight transporters in the United States. In 2017, there was a truck driver shortage of approximately 50,000. The situation worsened during the 2020 pandemic as truck freighters lost 63,000 jobs, six percent of its pre-pandemic number of truck drivers (Cheung, 2021). A shortage of truck drivers pushes truck companies to raise salaries to attract good drivers. A lasting solution to this problem could be a welcome relief to many U.S. truckers. One of the promising answers is to develop autonomous trucks.

It may take some time before any car company produces a fully self-driving truck. One of the vehicle companies driving the development of autonomous vehicles is Tesla. This electric carmaker is currently hard at work refining its Autopilot system, and as explained above, is on version 10 of its FSD for passenger cars. Truckers should take to heart the fact that, in 2016, an autonomous truck traveled from Fort Collins to Colorado Springs, covering 120 miles. However, that truck achieved this feat with a professional driver's assistance.

A lot is happening in the world of self-driving trucks as more companies are emerging. Many companies are testing self-driving technology, while government officials are working tirelessly to promulgate the necessary laws and regulations to ensure the safety of road users and truck drivers. Tesla introduced its first truck in 2017. It was a truck equipped with a radar system that can navigate the road in an emergency without a human driver. We'll begin to see Tesla trucks on the road when the battery supply problem is solved. Tesla is getting the chip-supply issue under control after developing its D1 chip, the company announced during the Tesla AI Day in August 2021. With Tesla's tendency to shift its deadlines, it's not easy to tell when its Semi will hit the American roads.

The good news is that companies like Walmart Canada are seeing the value of Tesla trucks. That company has ordered 130 Tesla Semi trucks, an investment worth around $23 million at $180,000 per truck.

Bezos Takes on Tesla With Amazon Zoox

Tesla is not alone in the quest for manufacturing autonomous vehicles. Bezos' Amazon has acquired Zoox and Aurora to play a major role in the transportation industry. Zoox is a California-based maker of autonomous ride-hailing vehicles. Amazon bought Zoox for two reasons—to enter the electric-vehicle race and to help this ride-hailing company bring its dream of producing autonomous vehicles to a reality. One of Amazon's strengths is long-term thinking, and autonomous vehicles is a long term effort. Also, it gives Amazon time to inject capital into this company so that it can materialize its business ideas.

In 2012, Amazon began a quest to automate its warehouse when it bought Kiva, and so now its attention shifts to ground-transport automation. The last mile is an important element of efficient delivery operations—a hard and expensive part of package delivery pound for pound. Eliminating human labor can lead to a drastic cost reduction for Amazon. Buying Zoox and winning the race to autonomous vehicles should aid Amazon in driving operational efficiencies and scaling. It's unlikely that Amazon wants to compete with ride-sharing companies, such as Uber and Lyft.

Following acquisition by Amazon, it took until December 2020 for Zoox to unveil its first robotaxi vehicle. Unlike other autonomous car designers, Zoox chose to build their unique robotaxi vehicle from the ground up instead of modifying existing cars. The car can accommodate up to four passengers, travel up to 75 miles per hour and travel for 16 hours on a single charge of 133 kWh. It's a small car that steers with all four wheels, making it easy to navigate it into tiny spaces. Zoox's autonomous car is equipped with an airbag system for safety purposes.

The four corners of the roof are equipped with rotating LiDAR units that provide an overlapping 270-degree field of view and are combined with radar and camera equipment that provide a 360-degree viewing range of 500 feet all around it. For autonomy, this vehicle utilizes artificial intelligence coupled with GPS data to identify objects perceived by the sensor suite. This way, it can formulate specific routes around obstacles, construction sites, and traffic congestion

The feeling you get when riding in this car is like a social setting, where passengers sit facing one another. It has sliding doors for easy access and a boxy shape. Here are two more unique features of Zoox's robotaxi:

- It has a four-wheel independent suspension to smooth out passenger rides; and
- Its airbags are integrated into the four seats with enclosing walls for added crash safety. Therefore, all passengers get full protection, as opposed to safer front seats in most cars.

Before Amazon acquired Zoox, Tesla had filed a lawsuit against the autonomous carmaker. Some employees had left Tesla for Zoox, and Tesla alleged that its former employees stole trade secrets for the benefit of their new employer, Zoox. Initially, Zoox denied that the said employers used Tesla's trade secrets to benefit it. However, in 2020, Zoox admitted to the allegations and settled the lawsuit by paying an undisclosed amount. Furthermore, Zoox had to undergo an audit to ascertain that the implicated employees were using Tesla's secret information.

Here's a statement from Zoox, acknowledging Tesla's allegation (Reuters Staff, 2020): "Zoox acknowledges that certain of its new hires from Tesla were in possession of Tesla documents pertaining to shipping, receiving, and warehouse procedures when they joined Zoox's logistics team." You can only hope that there's no tension that develops between Tesla and Amazon

Zoox as the race to develop autonomous vehicles heats up.

Discover Why Amazon Invested in Aurora

Prior to acquiring Zoox, Amazon invested heavily in car self-driving-technology maker, Aurora, in early 2019. Amazon wasn't the sole investor at the time, as Sequoia and the investment arm of Shell also put money in the startup. In total, Aurora received $530 million from the three aforementioned investors. The involvement of Amazon in the self-driving car is a clear sign that it wants to be a significant player, not just a consumer. Amazon's foray into driverless technology comes after the Seattle-based online retailer formed a team in about 2016 to focus on driverless-car technology. The team wasn't formed to build autonomous vehicles, but to act as the brain trust in figuring out how Amazon could leverage driverless cars. Amazon's intention is to move toward owning the bulk of its transportation needs and shipping chain logistics.

Amazon hinted on why it invested in Aurora by saying:

> We are always looking to invest in innovative, customer-obsessed companies, and Aurora is just that. Autonomous technology has the potential to help make the jobs of our employees and partners safer and more productive, whether it's in a fulfillment center

> or on the road, and we're excited about the possibilities. (Haselton & Bosa, 2019)

This complements the argument that Amazon wants to take much of the heavy lifting off of its human labor and thereby reduce transportation costs. As a result, this trillion-dollar company should become more profitable over time.

Aurora was started by former employees of self-driving car companies. It was co-founded by Drew Banell, Chris Urmson, and Sterling Anderson, some of the most experienced people on self-driving technology. CEO Chris Urmson is a former chief technology officer at Alphabet, parent of Waymo; Sterling Anderson is chief product officer and is a former design and launch lead of Tesla Model X; and Drew Bagnell is chief technology officer who worked for Uber's Advanced Technology Center.

Unlike Zoox and Tesla, Aurora doesn't manufacture cars, but partners with truck and carmakers. Among its current partners are major truck manufacturing players, Volvo and PACCAR, and top carmakers, Toyota, Volkswagen, and Hyundai. The core product that Aurora sells is Aurora Driver, and offers this as a service across trucking, local goods transportation, and ride-hailing. In the trucking industry, Aurora Driver will facilitate high-volume transportation of goods for shippers and logistic businesses between homes, businesses, ports, and depots. Aurora expects to launch its truck self-driving technology in late 2023 and then expand into other markets, including ride-hailing and last-mile delivery.

Machine learning is a critical element of most autonomous software, including for Aurora; however, more is required. For accurate representation of traffic situations, Aurora combines modern machine learning with powerful engineering, including in the areas of decision-making, perception, and state estimation. Aurora's engineers are focusing on level 4 autonomy, which can be applied in several ways.

In December 2020, Aurora announced that it was buying Uber's self-driving unit Advanced Technologies Group (ATG). This move not only adds self-driving expertise to Aurora, but brings technology to help Aurora drive toward its mission with verve. Uber sold its self-driving unit following a rocky path filled with problems such as a deadly accident and allegations of using stolen trade secrets. In the same announcement, Aurora also revealed that it is forming a strategic partnership with Uber. The partnership will enable Aurora to widely distribute its Aurora Driver, especially in the ride-hailing segment of the market. Uber invested $400 million in Aurora to support the partnership, and Uber CEO Dara Khosrowshahi joined Aurora's board of directors.

Aurora expects to change its name to Aurora Innovation following a business combination with Reinvent Technology Partners Y, a Nasdaq-listed special-purpose acquisition company. Furthermore, this new company will also become a public company listed on Nasdaq. A combination of acquisitions, investments, and business partnerships is moving Aurora forward toward realizing its mission.

Chapter 6:

Bezos Battles Musk in the Electric Vehicles Race; Can He Win This Race?

"There are really two things that have to occur in order for a new technology to be affordable to the mass market. One is you need economies of scale. The other is you need to iterate on the design. You need to go through a few versions." — Elon Musk

The electric-vehicle market, although young, is growing at tremendous rates. In 2020, it was valued at $171.26 billion, and it is forecast to grow at a compound annual growth rate (CAGR) of about 27.19% from 2021 to 2026, reaching $725.14 billion (Mordor Intelligence, n.d.). Factors responsible for this growth include rising costs of fuel and environmental concerns, especially the impact of global warming. Governments and companies around the world are increasingly concerned about the environment, and many will lead the way toward the usage of renewable energy resources. To cater to the increase in global demand for electric vehicles, there'll be a need for charging stations and various vehicles. Amazon and Tesla will be among some of the companies driving the adoption of electric vehicles.

The rivalry between Jeff Bezos and Elon Musk extends from space to Earth, from satellite Internet to freight transportation. Amazon invested in Rivian to help grow this Robert 'R.J.' Scaringe-founded electric carmaker. By doing this, Jeff Bezos goes head-to-head with Elon Musk in manufacturing electric vehicles. It's estimated that both Amazon and Ford Motor Company injected a total of $1.7 billion into Rivian, with $700 million coming from Amazon (Kolakowski, 2019). Rivian believes there's a more responsible way to explore and enjoy our world.

What Is Amazon Rivian?

Rivian is an electric-vehicle startup company that has been developing electric cars since 2009, and thus, officially became a Tesla direct competitor. This EV carmaker develops SUVs and pickups, and it may go public sometime in 2021. Currently, Rivian's valuation could be about $70 billion post listing on a stock exchange. The vehicles that Rivian is building are capable of covering 150 miles per charge. Rivian employs more than 3,600 employees in its Illinois manufacturing facility, plus its offices in Michigan and California, and also has more than a thousand job openings.

Amazon signed The Climate Pledge in 2019, in which it committed to become carbon neutral by 2040. To this effect, Amazon ordered 100,000 Rivian vehicles in 2019 and expected the company to start delivering customer packages in 2021 (Amazon, 2019). Amazon foresees having 10,000 of its ordered batch of electric vehicles on the road by 2022, with the balance coming in 2030. When that happens, Amazon hopes to reduce carbon emissions by 4 million metric tons annually by 2030 (Amazon, 2019).

The electric vehicles that Rivian builds for Amazon are delivery vans, and Amazon unveiled their look in October 2020. With the Amazon order alone, Rivian could make roughly $4 billion. The first van seen on the road was in a Los Angeles neighborhood, and it's identical to what Amazon unveiled in October. An Amazon manager and a Rivian engineer seemed to have fun in the new van. It's large and boxy and has large doors for easy loading and unloading packages from the optimized package space. The Amazon log was visible from all angles and sports a massive windscreen for enhanced visibility. Furthermore, the Rivian van had a manufacturer badge, indicating that it was from Michigan.

R.J. Scaringe, Rivian boss, pointed out that Amazon's van is an example of what they're trying to make. Although Amazon has previously worked with electric truck manufacturers, Scaringe hastened to assure Rivian customers that he and his team are building unique products. One of the unique features of the van is that, since it is an electric vehicle and is much quieter than a gasoline-combustible vehicle, it made a slight noise, for added safety, so pedestrians could hear it.

Amazon vans are fully connected to enable monitoring the speed, state of charge, and location of each vehicle in the fleet. This connectivity will come in handy to monitor the health of the vehicle and to reach out to users when repairs or maintenance are needed. Each Amazon van will sport a basic electrical and network structure and will have the same battery packs as the Rivian R1T, a pickup vehicle. One of the challenges in the last-mile delivery industry is high driver turnover. Therefore, making the Amazon vans comfortable was an important consideration.

The interior of the Amazon van is driver-friendly, featuring a climate-control system, large cargo space, touchscreen infotainment, a generous list of safety features, and top specification lighting. Rivian simplified the design of the Amazon van by speaking directly with Amazon drivers to figure out precisely what they wanted. That research resulted in optimized entry steps, which prevent slipping when wet and handles that are easy to use while handling packages.

While the R1T utilizes an aluminum skateboard framework, the Amazon van uses a steel housing and ladder frame. This difference is considerable and allows the two vehicles to be assembled separately. More importantly, the van has a large cabin space for moving around easily in the vehicle, and the driver's side doors are reinforced for added safety.

Amazon wrote on its blog in February that it, together with Rivian, has begun testing delivery vans four months prior to making customer deliveries. This is the normal part of the development process to ensure vehicles perform as expected. In these tests, Rivian wanted to check the vehicles' performance, including checking their safety durability in various climates for engineers to refine them. Amazon customers will be glad to know that the company is preparing buildings to house their new fleet of vehicles. Part of the preparations include installing thousands of charging stations in both North America and Europe.

Amazon Electric Truck for the Middle Mile

In September 2020, Lion Electric revealed that it will deliver 10 battery electric trucks to Amazon. The purpose of the trucks is to deliver customer packages for Amazon's middle mile. This deal is an extension of the order that Amazon made to Rivian for electric vans to move it toward its 2040 decarbonizing goal. Lion Electric, a Canadian electric vehicle company, announced in late 2020 that it has merged with SPAC Northern Genesis Acquisition Corp in a deal worth $500 million. Furthermore, Lion Electric announced that it will list on the New York Stock Exchange.

Lion Electric's deal with Amazon involves producing about 2,500 Lion 6 and Lion 8 electric trucks by 2025. This deal is exciting and could result in Amazon buying a stake in the electric company for up to 19.98% if they buy products worth $1.1 billion. Not only will Lion Electric produce electric trucks for Amazon, but it will also maintain the vehicles and train their staff. This deal between Amazon and Lion Electric started when the latter's CEO and founder, Marc Bedard, was a guest speaker at a trucking event. That's where Bedard met and invited Amazon's representatives to Quebec. There, Amazon audited and tested Lion's electric trucks, and the rest is history.

Lion's electric trucks are available in various setups, including refuse trucks, straight trucks, and reefer trucks. Customers can purchase Lion's trucks that are vehicle-to-grid-enabled and can cover up to 250 miles with a single charge. Amazon must have had confidence in Lion, considering that this electric company's vehicles have recorded more than 6 million miles in the 10-year life of the enterprise. Commenting on the deal with Amazon, Bedard said:

> This vehicle delivery for Amazon represents a very significant milestone for Lion. Amazon is a leader in adopting decarbonizing technologies that can improve sustainability among their trucking fleet. Our all-electric trucks will be a valuable addition to Amazon's trucking operations as they work to deliver on their sustainability goals. (The Lion Electric Co., 2020)

Tesla: The Top Dog of Electric Car Manufacturers

If you've been following electrical cars, you'd know that Tesla is by far the biggest battery-operated carmaker. Not only is Tesla a leading electric-vehicle carmaker, but it is also the fastest-growing brand, globally. In 2020 alone, Tesla delivered more than 500,000 cars, with the Model 3 being the best-selling electric vehicle. Nearly 90% of the delivered vehicles were Model 3 and Model Y, the models with which Tesla is appealing to the broader market (Carlier, 2021). At this point, Tesla has four production cars, Model S, Model Y, Model X, and Model 3. Tesla Model 3 and Model S are sedan versions, while Model X is an SUV and Model Y competes in the crossover-SUV segment.

Tesla is also in the process of developing a class 8 heavy-duty truck called the Tesla Semi. This truck features four independent electric motors on rear axles for movement and has an enhanced Autopilot for added safety. Compared to diesel trucks, the Tesla Semi can save logistics providers more than $200,000 in fuel. It does this because it requires half the energy costs needed to operate a diesel truck. The Tesla Semi consumes less than 125 kWh per 60 miles of traveling (Tesla, n.d.). Most importantly, the Tesla Semi will carry a competitive payload compared to similar diesel trucks.

Tesla's Upcoming Electric Pickup: The Cybertruck

Tesla is dominating the passenger car segment in the electric-vehicle race; however, this California-based company is also building battery-powered vehicles in other car segments. One of the segments Tesla will be joining is the light-duty truck segment with its Cybertruck, unveiled in November 2019. Tesla plans to sell the single-motor rear-wheel drive Cybertruck for $39,000, the dual-motor all-wheel-drive for $49,000, and the tri-motor all-wheel drive for $69,000. The single-motor version is expected to start production in late 2022, while the all-wheel drive version's production is slated to start in late 2021.

In the recent Quarter two 2021 Tesla earnings call, a question was asked about when the Cybertruck would be released. Tesla, through Lars Moray, vice president of vehicle engineering, said that it's finished with basic engineering and is now working on how the car is to be made. However, Moray stated that production of the Model Y takes precedence over making the Cybertruck. Customers could still see the Cybertruck going into beta phase close to the end of 2021. Ramping up production of the Cybertruck will take place at Texas only once the Model Y is in full production.

Musk reiterated a few times during the call about the difficulty of manufacturing a car as opposed to making a prototype. The biggest challenges Tesla is facing with scaling the Cybertruck and Tesla Semi are chip shortages and cell capacity. Musk admitted that they can produce the Cybertruck and Tesla Semi at a limited rate, and this will shoot up the cost of production per vehicle. When production of the Cybertruck happens, that vehicle will cover 500 miles per charge and tow up to an astonishing 14,000 pounds.

Is the Tesla Van in the Cards?

There are theories making the rounds that Tesla will manufacture a Tesla van in the future. Knowing Elon Musk, something like that is possible. Musk shared his master plan for Tesla on a blog and indicated that his electric company's plan went like this:

- Build a low volume and expensive car;
- Use the proceeds from the first car to create a cheaper, medium-volume vehicle; and
- Utilize the money from the second car to develop a high-volume, affordable car.

Musk went on to say that besides producing consumer vehicles, the electric car market needs heavy-duty trucks and compact passenger urban transport. Nowhere in the master plan does Elon Musk explicitly state that his company will build an electric van. However, there have been developments since then that suggest Tesla may create an electric van, however, Musk's master plan concludes by saying, "Expand the electric vehicle product line to address all major segments" (Musk, 2016). Could this expansion also include a Tesla van?

The Boring Company's Tunnel to California Airport and Electric Vans

On June 3, 2020, a San Bernardino County transportation agency, supported by the county's Transportation Authority's Board of Directors, approved The Boring Company's proposal to excavate a tunnel connecting Ontario International Airport with Rancho Cucamonga. The idea was seen as cheaper and faster than surface rail projects the county was considering.

Interestingly, electric cars are the proposed mode of transport to ferry passengers from one side to the other. It would take about 90 seconds to two minutes to complete the journey from Cucamonga to the airport or vice versa. The proposal initially called for the use of special Tesla cars, but one member of the board revealed that The Boring Company was working with Tesla to build electric vans. Such vans would seat up to 12 passengers and their luggage, and they would increase the capacity to 1,200 people a day. Elon Musk has yet to confirm if Tesla would build electric vans for the San Bernardino County tunnel.

Prior to the above deal, there had been talk of Musk creating an electric van. Those conversations were triggered by a potential collaboration on an electric van between Tesla and Daimler Mercedes-Benz in late 2018 and early 2019. The possible collaboration story was triggered by a tweet from a certain @ScaleyMurrsuit who said, "Shame that's not a Tesla VAN. Because we REALLY REALLY REALLY REALLY need a good tough as nails electric trades van. We burn fuel like crazy" (ScaleyMurrsuit, 2018). In reply, Musk said, "Maybe interesting to work with Daimler/Mercedes on an electric Sprinter. That's a great van. We will inquire" (Musk, 2018).

When this talk of collaboration was taking place, Daimler Mercedes-Benz had already unveiled their electric version of their famous Sprinter in February 2018. True to their forecast, Daimler Mercedes-Benz began production of its first eSprinter in December 2019. Employees and invited guests graced an event to celebrate the maiden production of the Mercedes-Benz eSprinter. This car has top-spec features such as reverse camera, electrical parking brake, blind spot assist, and active brake assist.

This van is developed for urban mobility and has the same loading volume as its internal-combustion-engine Sprinter. Customers can customize their vans based on payload and battery power, available in three or four battery units. Battery choice determines the range the car drives before needing a charge, and you can charge the battery to 80% full in 30 minutes. Mercedes-Benz had sold 1,200 eSprinters to Amazon when the two companies signed The Climate Pledge. To make the total that Amazon ordered, Mercedes-Benz will produce 600 midsize electric vans called the eVitos.

Yet, while Daimler Mercedes-Benz was progressing with development of their eSprinter, former CEO Dieter Zetsche confirmed that Tesla had shown interest in a collaboration on an electric van. The talks were made possible by Jerome Guillen, a former Tesla executive who had previously worked for Daimler. Since then, there has been little talk of this collaboration. However, Musk hasn't ruled out the possibility of Tesla building an electric van in the future.

During Tesla's quarter four and financial year 2020 earnings call, RBC Capital Markets asked Musk about the electric van market. With no hesitation, Musk confirmed that Tesla will make an electric van in the future, but didn't commit to a date. Most importantly, Musk revealed that the biggest constraint to producing an electric van is battery supply, a factor that influences the production of the Tesla Semi.

A possible solution to this challenge is utilizing solar panels. This possibility surfaced when Elon Musk appeared in a Joe Rogan, a comedian, podcast. Musk had said that using solar panels on passenger and luxury cars is not effective because the surface area of their roofs are small. In contrast, a van's roof is a large surface that can be enough to power it. Furthermore, Musk indicated that you could extend the van's roof for charging the vehicle when it's stationary. Despite renewed interest in the electric van, Tesla has not yet confirmed that it is going to build it. Perhaps that's because the company currently has a lot on its plate.

Chapter 7:

How SpaceX Plans to Transport Cargo by Air and What Bezos Is Doing About it

*"If you're not flexible, you'll pound your head against the wall and you won't
see a different solution to a problem you're trying to solve."* —
Jeff Bezos

It appears the word 'impossible' does not exist in Elon Musk and Jeff Bezos' dictionaries. These two billionaires keep dreaming about the future, and it seems exciting to them. What's powerful is that they're able to share their dreams and attract like-minded people to take on what may appear to be pipe dreams. There's no better industry to illustrate what's possible for these two entrepreneurs. In this chapter, you'll discover how each of these top two richest persons in the world are dreaming, when it comes to air transportation.

SpaceX Continent-to-Continent Rocket Cargo Transportation

On September 29, 2017, SpaceX CEO Elon Musk unveiled revised plans to travel to Mars and the Moon at a space-industry conference. There, the unexpected happened: Musk told his audience that SpaceX would, in some future time, use the same interplanetary rocket system for long-distance travel here on Earth. Using this system, Musk estimated it would take 30 minutes for passengers to go anywhere on Earth at a cost equal to that of an economy airline ticket using Starship, a mega-rocket.

However, the developments at SpaceX and other space exploration companies have caught the attention of the U.S. government. The U.S. Transportation Command (USTRANSCOM) announced on October 7th, 2020, that it is partnering with SpaceX and Exploration Architecture Corporation (XArc) to investigate the possibility of rapid transportation of time-sensitive goods through space. The advantages of such a transport mechanism will allow for the rapid movement of critical resources without the need for refueling or stops. This could be one of the greatest transportation achievements of our time.

Following its work with SpaceX and XArc, the U.S. Air Force requested nearly $50 million to fund further research into this program, called Rocket Cargo. The aim is to develop a 100-ton rocket fitted to airdrop cargo anywhere on Earth. Currently, the Falcon 9 can carry about a 50,700-pound payload into a low orbit above Earth, which is a far cry from the weight that the military is looking to transport. It's unclear, however, how much weight and altitude the Falcon 9 can reach for point-to-point transportation of cargo on Earth.

When required, a Rocket Cargo would launch within minutes from an established base in the United States, such as the Kennedy Space Center on the East Coast. Within a short time, the rocket would reach its targeted drop zone and separate from the cargo capsule. The capsule would descend into a landing zone and then land. Here, loadmasters would unload the capsule and hand over the load to users.

Musk Dreaming About a Supersonic Jet

Elon Musk tweeted this on October 25, 2020, in response to a World of Engineering tweet: "Sigh…there should be a new supersonic jet, this time electric." The original tweet was in commemoration of Concorde's last flight before being retired. This plane took under three hours to fly from New York to London because it could reach a cruising speed of 1,350 mph. Musk's tweet suggests he still had a dream of seeing an electric plane someday.

What stands in his way of developing an electric plane is the unavailability of a suitable-sized battery. While the Tesla Model 3 has a battery that delivers around 250 watt-hours per kilogram of energy density, an electric vertical takeoff-and-landing vehicle could require at least 400 watt-hours per kilogram. The good news is that Tesla is working hard to produce its own batteries, which are estimated to increase the battery range by 54% for electric cars and drop the price per kilowatt-hour by 56%. Although increasing the battery density is geared for cars, the battery size might be enough for an electric plane's takeoff.

In August 2020, Tesla's battery research partner, Jeff Dahn, from Dalhousie University, updated those interested in battery technology on a new approach to produce high-energy density cells. Such batteries could be used to power electric cars, electric aircraft, and drones. In summarizing their research findings, Louli *et al.* (2020) said,

> Anode-free lithium metal cells store 60% more energy per volume than conventional lithium-ion cells. Such high energy density can increase the range of electric vehicles by approximately 280 km or even enable electrified urban aviation. However, these cells tend to experience rapid capacity loss and short cycle life.

Although such a battery has a short lifecycle, the researchers in the above-mentioned study were able to increase the battery's life to 200 life cycles. This progress is probably inadequate because 800 to 1,000 cycles seem to be the norm for commercialization. Like all development work, this research paves the way for production of batteries capable of fully replacing traditional ones and enabling transportation that utilizes electricity.

If renewable energy is the way to go for Musk and his companies, it's understandable why his focus has been on Tesla cars than on an electric plane. Aviation accounts for about two percent of global carbon dioxide emissions, while transport contributes 24%, of which 72% is from road transport (Brown, 2019).

What Amazon Is Doing in the Air-Transportation Space

Why Amazon Introduced a Fleet of Cargo Aircraft

Operating an air service is expensive, but Amazon is taking it on and growing its package delivery fleet. For cost-effective air service, Amazon will need to have enough inventory to fill its planes. The 2020 pandemic helped drive an increase in online sales, an important element for Amazon's air-package-delivery strategy. Amazon Air, a freight transport arm of the giant online retailer, transports Amazon packages. In late 2015, Amazon Air, codenamed Project Aerosmith, launched a test cargo operation from Wilmington Air Park. At that time, Amazon's air-freight operation was called Prime Air, hence, its planes were branded "Prime Air." Then, in 2017, Amazon changed its air-fleet branding to Amazon Air to differentiate it from its new drone-delivery service, called Amazon Prime Air. However, the Prime Air logo still appears on Amazon planes, which can be confusing.

Previously, Amazon Air had leased its aircraft; however, the company recently bought 11 aircraft from WestJet and Delta Air Lines to be converted into cargo planes. By purchasing these aircraft, Amazon is becoming a full-service carrier, with its own fleet of delivery vehicles that carry packages around the world. Amazon has opened a European hub, called the Leipzig hub, to expand its international shipping network. To speed up delivery for Prime customers, Amazon packages will flow through this cargo facility and will eliminate the need for third-party parcel carriers. Amazon Air leased two Boeing 737-800 aircraft and hired ASL Ireland Airlines to initially operate the two daily flights.

To support supply chains across the Pacific, Amazon seems to rely on chartered flights and other contractual agreements. For this reason, it doesn't need to keep numerous trans-Pacific flights. Transoceanic off-season international flights have been reduced to semi-regular return flights from Amsterdam to Chicago. To support its air-transportation strategy, Amazon has set up gateway facilities in Riverside, Calif.; Alliance, Texas; and Wilmington, N.C. These facilities have dedicated parking spaces to enable unloading Amazon packages at the ramp to continue Amazon delivery on the last mile. Aircraft are essential to move goods and parcels rapidly between warehouses and fulfillment centers. Shipping connects manufacturers to their warehouses, while conventional cargo carriers use large aircraft from their facilities, which appear more suited to the task.

Amazon's aircraft logistics is an achievement that would not have been possible without talented people—both technical, non-technical—and the corporate team, who invented its global operations. The primary reason for this innovation is to serve its growing number of Prime customers who love fast delivery, low prices, and the wide choice that Amazon offers.

What Is Amazon Drone-Delivery Service?

Amazon intends to deliver up to five pounds of customer packages within 30 minutes through its Prime Air drone deliveries. Before this system could fully operate, Amazon must satisfy the FAA that its drone operations will be safe. One of the important steps that Amazon took was to obtain a Part 135 Air Carrier Certificate, which it did in August 2020. This is a sign that the FAA has confidence in Amazon's safety procedures and autonomous drone-delivery operations. FAA approval of Amazon Prime Air delivery is a promising step toward future innovative parcel-delivery services and a big step forward for e-commerce. We will see more green lights, in the future, to enable more companies to provide drone-delivery services as the FAA streamlines its approval process and adapts its rules to current technology.

The development of autonomous drone technology has recently grown rapidly. New rules for parcel delivery and last-mile logistics are expected for the commercial drone industry in the United States. In 2019, the Amazon drone, MK27, was unveiled at re:Mars, which was the online retailer's global AI event on Machine Learning, Automation, Robotics, and Space, in Las Vegas. The drone has a hexagonal frame, and it can take off and land like a helicopter. Once in the air, it tilts like a regular airplane for more efficiency. It has several sensors and computer systems that allow it to touch down at home without hitting power lines that could pose a danger to humans and pets. Amazon isn't willing to delve into the details of what the actual hardware on board will look like.

Gur Kimchi, Amazon's VP for its Prime Air program, said in 2019 that the system will use more than one operating system and one CPU architecture. But, it is the integration of sensors, AI, and smarts into the actual design of the drone that will make it work. The autopilot uses sensor data developed by Amazon itself and gives the drone six degrees of freedom to reach its destination. The angle of the box in the middle of the drone houses most of its smarts, so the packages it delivers don't have to pivot. Amazon, for its part, says that certain items will be available only with drones. Not all items on its website will offer two- or one-day shipping, and even fewer items could be shipped to a customer within two hours.

Amazon is not the only company looking to expand commercial-drone delivery. In April 2019, Wing, owned by Alphabet, founded in 2012, became the first drone-delivery company to receive FAA approval for commercial delivery in the United States, and UPS received FAA approval to operate a fleet of drones as an airline last October. The small drone is designed to deliver small packages weighing 3.3 pounds or lighter. The drone hovers about 20 feet above the ground and lowers packages by attaching a leash to the ground. Customers do not have to interact with the drone. The company also offers the drone with an OpenSky navigation system for parcel delivery.

Amazon and other companies that want to revolutionize the retail trade through drone-deliveries have made significant strides in recent years. They have invented new devices that have shown, to a limited extent, at least, that they are capable of flying long distances and carrying the payloads required for packages. Routine deliveries, however, are likely to take years.

Various parts of the U.S. government have expressed concern about the misuse of drones, a concern heightened by a series of incidents in 2018 in which a drone caused the closure of London's Gatwick Airport for 33 hours. Amazon's drone-delivery test, however, shows the technology isn't there yet. Consumers are also concerned about the privacy of drones that use GPS and cameras to find homes and deliver packages. And, delivery-service providers could lose their jobs due to automation.

Amazon claims that its traffic-management system is easier to use because the different operators are in the same airspace, and the system is connected to the Internet. For added safety, drones can fly at low altitudes, as low as 400 feet. Unlike roads with fixed routes, there are many more ways to fly from point A to point B, and the navigation of a drone in the air is very different from that of a car driving on the road.

Chapter 8:

Who Will Win the Tightly Contested Freight-Transportation Race?

"I think it's very important to have a feedback loop, where you're constantly thinking about what you've done and how you could be doing it better." — Elon Musk

Freight transportation is another industry in which both Elon Musk and Jeff Bezos are actively participating. They may not compete directly against each other; however, they're both impacting segments of the freight industry they each operate in. The strategies that each employs could dominate the freight-transportation industry in the long term.

Amazon Extends its Tentacles Into Freight Transportation

Amazon Enters the Freight-Forwarding Business

Amazon, through its China arm, became an ocean-freight forwarder in 2016, a move that was intended to reduce shipping costs and potentially serve as a logistics brokerage. However, Amazon didn't want to operate ships; instead, it wanted to subcontract the work. Freight forwarders are known as "Non-Vessel-Operating Common Carriers" (NVOC), in the ocean-shipping industry. These companies buy space on ships and find cargo to fill them. NVOCs like Amazon rent out space on steamship lines and sell it to their customers, the way Amazon leases cargo planes. This move paved the way for Chinese manufacturers to sell their wares in the United States. More importantly, consumers don't have to worry about lost parcels as they change hands between transporters.

By October 2019, Amazon had a total of 350 intermodal containers, and its website listed around 779 job openings in its transport and logistics sector (xChange, 2019). This was a clear sign that Amazon wanted to play big in its supply chain and, possibly, to attain more control so that its business could become profitable. This Amazon service didn't just happen in 2016; Amazon had been testing maritime waters for at least three years. The new logistics service for Chinese exporters, ocean freight, Amazon's air freight, labeling, sorting, trucking, and distribution allows Amazon to capture the growing business of shipping freight from its own Fulfillment by Amazon (FBA) warehouses. Furthermore, the Chinese can market directly to consumers and compete with other online and traditional retailers.

Amazon has not penetrated deep into the logistics world with dedicated platforms, wire-by-wire portals, and simple operations. The company's greatest added value for ocean freight is price. A seller's decision to use this Amazon service doesn't affect the selling price of their goods. Instead, sellers are paid less for their products because Amazon handles the shipping. A seller arranges delivery from China to an Amazon warehouse by telling their freight forwarder the address of the Amazon warehouse. In turn, Amazon charges the seller a delivery fee based on terms you agree on with your supplier's Free On Board (FOB) Shanghai or Ex Works (EXW) Shenzhen. This service helps sellers shorten their delivery time. Shipping your containers to your own warehouse at third-party logistics (3PL) can extend your transit times. In contrast, using Amazon freight-forwarding services reduces the contact time between your products and containers.

Amazon in the Freight Brokerage Space

Amazon took the first step toward entering the freight-brokerage business since 2018 when it tested the market in five states. What it did was it matched truck drivers with shippers through its arm called Amazon Freight. This online freight service enables Amazon to better manage its existing network of carriers and speed up the process of cargo matching. Here's what Amazon says about its current freight business:

> We blend advanced technology with a network of 30,000+ Amazon trailers and carriers to move your full truckload freight—simply and reliably. Put the power of Amazon behind your shipments, with a partner that's here to meet the demands of today and help you navigate the road ahead. (Amazon Freight, n.d.)

Amazon Freight competes against such companies as Uber Freight, XPO Logistics, and C.H. Robinson. Uber Freight launched in the U.S. in 2017, and it made $313 million in the fourth quarter of 2020, an 8.6% rise over the previous quarter. Its earnings before interest, tax, depreciation, and amortization (EBITDA) improved to a loss of $41 million compared to $55 million in 2018 (Kingston, 2021). C.H. Robinson, one of the largest logistics and freight-brokerage companies in the world, registered $16.2 billion in revenue in 2018, a 5.9% increase from 2019.

Amazon rolled out its freight brokerage throughout the United States in May 2020. As a result, more shippers can now access Amazon's freight services through its portal freight.amazon.com. Carson Krieg, co-founder of Convey, a tech firm that helps retailers with logistics and supply chain processes, believes Amazon wants to control the logistics market and supply chain to control its shipping costs.

The overall freight-trucking market in the United States is estimated to hit $515 billion in 2021. The U.S. accounts for nearly a quarter share of the global freight-trucking market. China is estimated to reach a market size of $587 billion by 2026, a CAGR of 6.3% (Global Industry Analysts, Inc., 2021). Countries such as Japan, Canada, and Germany are focused to grow by 2.5%, 4.7%, and 2.6%, respectively.

Introduces the Tesla Semi for Freight Transportation

Elon Musk publicly unveiled the Tesla Semi truck at a live-streamed event in Hawthorne, Calif. in November 2017. The truck was pitched as the safest and most comfortable truck ever developed. Musk unveiled the Tesla Semi as a fully electric Class 8 truck, which means it weighs 33,000 pounds and can haul 80,000 pounds of freight. The aerodynamic design of the truck has a drag coefficient of 0.36, slightly lower than that of the Bugatti Chiron supercar. Furthermore, the Tesla Semi, as Musk revealed, would cover 500 miles on a single charge and at maximum weight.

The Tesla Semi's weight includes the electric motors and transmission, the truck itself, battery pack, and powertrain. For charging, this vehicle utilizes existing Tesla charging systems and requires plugging into a minimum of four supercharger stalls to fully power its battery pack. Finding economies of scale for fast charging was a necessity to charge the Tesla Semi. At the unveiling of the Semi, Musk revealed that it would cost about $1.26 per mile to operate it, which is cheaper than the $1.51 per mile for the diesel truck (Cava, 2017).

Like all brands of vehicles, the price of a Tesla Semi depends on the model you buy. The lowest price you can pay is $150,000 for the 300-mile-range truck, while for $180,000 you can buy the 500-mile-range version. To own either model, you'll need to put down $20,000. There's also a third model called the "Founders series,'" which you can buy for $200,000.

When Musk introduced the Tesla Semi in 2017, he was targeting 2019 as the year of production and delivery of first trucks to customers. However, production issues of the Model 3 delayed the production of the Tesla Semi trucks. Even so, some organizations have already placed orders for semi trucks, such as the following companies (Manthey, 2020):

- UPS, which ordered 125 units;
- PepsiCo placed an order for 100;
- Sysco ordered 50;
- DHL has ordered 10;
- Walmart Canada ordered 130; and
- Anheuser-Busch has ordered 40.

In July 2021, Tesla advised that its Semi trucks will be available in 2022 as per the following message:

> We believe we remain on track to build our first Model Y vehicles in Berlin and Austin in 2021.

The pace of the respective production ramps will be influenced by the successful introduction of many new product and manufacturing technologies, ongoing supply chain-related challenges and regional permitting. To better focus on these factories, and due to the limited availability of battery cells and global supply chain challenges, we have shifted the launch of the Semi truck program to 2022. (Tesla, Inc., 2021)

The Boring Company Digs Freight-Transportation Tunnels

The Boring Company initially dug tunnels by using the Godot, which has become outdated. The next machine used could automate the concrete laying and utilized an Xbox controller to operate it. The present tunneling machine, Prufrock, can dig at a rate of one mile per week, which is six times faster than the previous generation machines (The Boring Company, n.d.). However, this is still slower than the rate at which a garden snail burrows the ground. Musk and his team at The Boring Company are not going to settle for slow tunneling speeds. Freight tunnels that The Boring Company digs could be used for both in loops for intra-city container transport. The passenger transport tunnels are a means to develop the tunneling technology and feasibility of cargo transportation.

Within two years of its inception, The Boring Company won contracts in Chicago, Los Angeles, and Hawthorne, Calif. The Boring Company isn't focused only on passenger-traffic tunnels, but also freight-transportation tunnels.

According to The Boring Company website, one of the products this company produces is a freight tunnel that's 12 feet in diameter, which barely fits a standard shipping container. Included in the scope of work will be permitting, CCTV video system, LED lighting, environmental review, and project engineering. You can imagine what an impact having freight tunnels could have on shipping in port areas such as the Port of Los Angeles (L.A.). This port is the largest in the Western Hemisphere, and it's a gateway to access a large part of North America's trade. What does this mean for freight transportation?

Up to this point, The Boring Company was aiming to build tunnels 12 feet wide. A standard shipping container is 8 feet wide, a requirement for it to be locked with others on a cargo vessel, and either 8 feet or 9 feet high. A tunnel that is 12 feet in diameter would barely be enough for a standard shipping company considering the carrier mechanism, which has to be less than 3 feet high. A recent article by Bloomberg revealed that The Boring Company is quietly gunning for 21-foot diameter tunnels, which could attract a wider customer base for them (McBride, 2021).

Tunnels that are 21 feet in diameter would be wide enough to accommodate two shipping containers side by side. According to the Bloomberg article mentioned above, The Boring Company is pitching potential customers for a tunnel that is 21 feet in diameter for shipping containers. One of the deck slides features three images showing shipping containers within tunnels. The first image shows a 12-foot-wide tunnel with one shipping container, and it's clear that the space is barely enough for the single container. For contrast, The Boring Company added a second image in which a single shipping container fits well within a tunnel that is 21 feet in diameter. To up the ante, The Boring Company shows, in the third image, how that 21-foot-wide tunnel can carry two shipping containers side by side, a setup they call a "dual carrier." In all the setups, the shipping containers appear to be traveling through a battery-operated freight carrier that has the same length and width as the shipping container, though slim.

It's not a question of whether it's possible to dig these types of tunnels, but a question of how much it would cost to do so. Furthermore, transporting freight through a channel should be a cheaper alternative to road transport. As explained in Chapter 1, digging wider tunnels increases the cost of tunneling. Since The Boring Company intends to widen its tunnels for freight transport, the tunneling cost is expected to rise. However, improved digging technology coupled with an efficient way of handling debris from tunneling could help keep costs down. The Boring Company has sold the idea of wider tunnels to California's San Bernardino county. If the project succeeds, it would help relieve road congestion around Ontario, Chino, and other nearby cities.

Inter-City Hyperloop Container Transport

There are various ways of transporting cargo from ports to warehouses across the United States. One more common method is by using intermodal transportation, in which more than one type of transport mode is utilized. For example, rail may be combined with trucks and trailers. A hyperloop is more situated for transporting goods between cities that are about 900 miles apart. Beyond that, air transportation might be more cost-effective, and therefore, a better alternative. One of the issues with this approach is that containers are handled more before they reach their destination.

This problem can be solved by using inter-city hyperloop container transport. Elon Musk was the first person to suggest usage of the hyperloop as an alternative mode of passenger transportation, as mentioned in Chapter 1. However, nothing prevents him from using the hyperloop to transport cargo between cities. A hyperloop is a low-pressure tube consisting of capsules moved at high speeds throughout the length of that tube. The difference between a passenger and cargo hyperloop would lie predominantly in the design of the capsule. A capsule is the actual component that holds the load being transported.

Although the hyperloop technology is still new, there has been progress made in the passenger-transportation space. In November 2020, Virgin Hyperloop One successfully accelerated two passengers for a short distance, reaching a top speed of 99 miles per hour (mph) before stopping. This experiment proved that the technology works. However, for inter-city container transport, there'll be a need to develop a suitable capsule and tube.

Chapter 9:

The Race to Dominate Global Satellite-Based Broadband Internet Service

"Failure and invention are inseparable twins. To invent, you have to experiment, and if you know in advance that it's going to work, it's not an experiment." — Jeff Bezos

SpaceX and Amazon are keen on making the Internet accessible to everyone, especially to people in rural areas. SpaceX's Starlink has already deployed more than 1,600 satellites to orbit and is beta-testing its Internet service in selected areas in Canada, the United Kingdom, and United States. However, Amazon's Project Kuiper has not yet sent any satellites into orbit, but plans to launch at least 3,200 satellites soon. There has been tensions between Amazon and SpaceX concerning the race to deliver satellite Internet to marginalized areas, especially to rural areas.

Why There's Need for Satellite Internet

There are certain areas throughout America and other countries that are sparsely populated. It's costly to install broadband infrastructure in those areas and could result in high prices simply for the access. Since few people would be able to afford such a service, many Internet service providers completely ignore installing broadband Internet in these areas. There are about 14.5 million Americans, 4.4% of the U.S. population, who don't have access to fixed Internet broadband at minimum required speeds, that is, broadband at 25 megabits-per-second download speed/3 megabits-per-second upload speed (Federal Communications Commission, 2021). Approximately 11.3 million of those people are in rural parts of the United States, which is an equivalent of 17% of the U.S. rural population (Federal Communications Commission, 2021).

The importance of access to reliable Internet became clear during the 2020 pandemic when people started working and schooling from home. This was a challenge that Project Kuiper and Starlink could help solve. Starlink had already run tests in several U.S. counties and found that speeds of about 70 Mbps were common. In certain cases, users were reaching speeds of 170 Mbps. Once the Starlink's satellite constellation is operating at optimum capacity, its customers should be able to access a fast and reliable Internet anywhere in the United States. Currently, a Starlink satellite dish costs $499, while, for continued service, you'll pay $99 monthly (O'Callaghan, 2020). For many people, these costs are steep, even with a $50 FCC Internet subscription subsidy.

Starlink has had a head start over Project Kuiper. However, both envisage spending around $10 billion for the infrastructure they need for the projects to succeed. Since both will be launching their satellites to low orbit, they'll likely deliver higher download and upload speeds than satellites for older companies. Since Starlink and Project Kuiper's are direct competitors, what happens with one may impact another's project.

The Clash of Amazon's Project Kuiper and SpaceX's Starlink

One of the major issues with satellite Internet is that satellites from different companies could interfere with each other. For example, Starlink's satellites may interfere with Kuiper's or other constellations from OneWeb and other operators. Not surprisingly, the FCC, when approving Project Kuiper, informed Amazon to avoid interfering with others' satellites. This could be a problem.

A few months prior to the FCC approval of Project Kuiper, Starlink had requested that the FCC approve their change of orbits for future satellites. Starlink argued that it wanted to improve service and simplify deorbiting of satellites. Starlink wanted to orbit their future satellites at 336 to 354 miles high, as opposed to their initially approved 684 to 823-mile height above the earth. On learning about this, Amazon wasn't happy and filed for an objection to the Starlink request. Other ventures also filed objections, arguing that the change of orbit by Starlink would put their own plans in disarray.

According to (Sheetz, 2021) on CNBC, Amazon said the following, regarding the changes that Starlink wanted to make:

> The facts are simple. We designed the Kuiper System to avoid interference with Starlink, and now SpaceX wants to change the design of its system. Those changes not only create a more dangerous environment for collisions in space, but they also increase radio interference for customers. Despite what SpaceX posts on Twitter, it is SpaceX's proposed changes that would hamstring competition among satellite systems. It is clearly in SpaceX's interest to smother competition in the cradle if they can, but it is certainly not in the public's interest.

In defending the modifications of his Starlink project, Musk tweeted this in response to a CNBC article: "It does not serve the public to hamstring Starlink today for an Amazon satellite system that is at best several years away from operation" (Musk, 2021). The question is whether the FCC will approve Starlink's request or not. If they do, will Amazon and other complainants agree with the FCC's decision? It's likely that if the FCC approves Starlink's request, other operators may seek the interventions of the courts.

Progress of Project Kuiper

Project Kuiper may have not yet deployed its satellites to orbit, but it's making progress. United Launch Alliance (ULA) announced on April 19, 2021, that it has partnered with Amazon to deploy satellites for their Project Kuiper. ULA provides its customers with reliable and innovative launch solutions, and they're excited to be a part of Project Kuiper, as it helps many more people all across the United States and worldwide to access the Internet.

The rocket to be used, Atlas V, will make nine launches for Project Kuiper. This rocket is known as the workhorse for space launch, having taken part in several missions for both government and commercial customers, including Mobile User Objective System (MUOS) constellations and Mars 2020 missions. Furthermore, ULA has successfully launched at least 140 missions for various purposes, such as tracking severe weather and GPS navigation, and to learn the complexities of the solar system.

In December 2020, Amazon completed its initial development of an antenna to connect users to the Amazon network. Tests of the Kuiper antenna have been promising because the antenna delivered Internet data at speeds of up to 400 Mbps. This is impressive considering that some of the tests included streaming 4K-quality video from satellites nearly 22,000 miles away from Earth. You can imagine the performance of this antenna when receiving data from 372 miles from the earth's surface!

Amazon took another important step toward achieving their Project Kuiper objectives when it acquired skilled employees experienced in Internet connectivity from Facebook. Facebook had built a team of Internet-connectivity specialists when it intended to expand Internet access to people in remote areas via satellites. Unfortunately, the company halted its attempts in 2018, and Amazon's Project Kuiper benefited from that.

Elon Musk Wants to Connect Starlink Satellite Internet to Moving Vehicles

As much as Project Kuiper is making some progress, Starlink is moving ahead fast. Originally intended to connect homes and offices to satellite Internet, Starlink now wants to hook up cars, jets, and ships to its network. In November 2020, SpaceX filed a request with the FCC to test one user terminal mounted on a maximum of five private jets. The tests involved checking performance while those private jets are in flight over the United States and when they are on the ground at an airport's tarmac.

A user terminal is a small ground device that connects to a satellite Internet network. Presumably, an aircraft-mounted terminal would be electrically similar to a ground one, except perhaps physically different. SpaceX wants to mount its Starlink user terminal on a jet for up to two years. Prior to filing a request to test its user terminals on jets, SpaceX had made a similar request to test on its seafaring ships.

SpaceX uses numerous ships to recover its rocket boosters, rocket nose cones, and capsules after its air missions. In its application to the FCC, SpaceX wanted to mount 10 user terminals onto two autonomous spaceport drone-ships and up to 10 seagoing platforms and run tests for up to two years.

In March 2021, SpaceX possibly realized that the individual filings they were making to the FCC regarding testing moving vehicles were too many. So, they decided to file a blanket request that covers all moving vehicles they had in mind, including jets, ships, cars, and trucks. The new filing didn't indicate how many user terminals would be tested because it isn't a requirement. User terminals for moving vehicles feature mountings different from those on the user devices for consumers.

Furthermore, the March filing considered the implications of Starlink in another country's airspace. So, SpaceX promised that when a U.S. aircraft leaves the United States, Starlink will operate by the FCC or that nation's rules, whichever is more stringent. It is increasingly becoming clear that SpaceX is interested in ensuring Internet connectivity while on the move, an important need for truck drivers, international flights, and European freighters. While Starlink user terminals are earmarked to possibly be installed on Tesla cars, Musk has not yet committed to it happening. The current mobile terminals are designed for aircraft, ships, and large trucks, and are too big for the Tesla cars.

Self-Driving Cars and Starlink

If you live in or visit rural America, you know how frustrating it can be to lose Internet connectivity and receive poor cell phone signals. The same would be the case for an autonomous car driving through such areas in the United States. Imagine an autonomous Tesla driving through Aurora County in South Dakota; when it gets to an area with either a poor or no Internet signal, the car will stop. How would you rescue yourself? That's why it's necessary to have strong Internet coverage throughout the United States if autonomous cars are the way to go. Morgan Stanley agree with this view in at least two ways (Fox, 2021):

- Fully self-driving cars may require satellite communications for uninterrupted operation, especially outside areas where 5G network is available; and
- Satellite communications offer triple-redundant safety and security, such as providing backup for navigation and connectivity.

Autonomous cars transfer terabytes of data between automobiles and AI-powered processors for seamless self-driving. For this reason, Starlink and Kuiper could play a major role in the production of autonomous vehicles because of their ultra-fast connections and low latency. With a self-driving car, you'd get in it, program your destination, sit back and let it take you there. That would free you to read a book, watch a movie, or work without worrying about you or your car's safety.

Tesla is evolving its suite of self-driving features faster than its competitors. Currently, Tesla has two systems that deliver different levels of automation: one is called Autopilot and the other is Full Self-Driving. Every Tesla you buy in 2021 has the Autopilot system enabled. This system boasts a traffic-aware cruise control that accelerates or slows down your car to sync with the speed of cars around. It also has a lane-keeping assistant system to ensure the car stays in between lane markings.

In contrast, fully self-driving vehicles are rated at a level 5 by the Society of Automotive Engineers (SAE), which means that such a car drives itself on any road and in any conditions. No company has yet produced a car that's fully self-driving, including Tesla. The Full Self-Driving system in Tesla cars is more sophisticated than the Autopilot system and is at SAE level 2, as mentioned in Chapter 5. It can change lanes on its own, park, and back out of a parking spot. Tesla is currently beta-testing an advanced Full Self-Driving system that can stop the car at red traffic lights, stop signs, and exit highway off-ramps. Importantly, you can't completely take your eye off the road because the system isn't autonomous yet. Although the Full Self-Driving system has not yet reached SAE level 5, the constant improvements Tesla is making could turn its cars into autonomous vehicles.

Terrestrial Internet connectivity is now moving toward 5G technology, which could reduce latency; however, these connections are rare in rural areas. Most importantly, cellular connectivity is not foolproof. As such, it's essential for companies like Tesla to move toward autonomous cars at the same time with initiatives that deliver high-speed Internet connectivity. Furthermore, it's important for autonomous vehicle carmakers to have control over their Internet network to provide security to minimize or prevent hacking. Tesla can tap into SpaceX's Starlink network for its autonomous vehicles to assist with that.

Tesla pioneered over-the-air (OTA) software that enables owners to update their vehicles' software. Currently, updates occur through cellular technology and could limit the kind of improvements carmakers could make due to slow data transfers. However, using satellite Internet, Tesla could push software updates to its cars quickly through SpaceX's Starlink network.

The thoughts of connecting cars with satellite Internet are not only in the minds of American carmakers. Geely, the largest Chinese private carmaker, obtained approval from China's National Development and Reform Commission (NDRC) to manufacture satellites for connectivity, navigation, and communication. These satellites will serve the same purpose as SpaceX's Starlink satellites and will be important for producing self-driving cars. Geely plans to produce 500 satellites a year. starting from October 2021, and will use the $326 million it had originally allocated for this project in March 2020.

Amazon's Project Kuiper could also consider being a middleman for getting satellite data to third-party carmakers and other technology startups. Amazon has ground stations that connect satellites with its web service setup. This puts Amazon at an advantage in that they don't have to build an extensive network of ground stations. Nothing seems likely to block Amazon from connecting its Kuiper satellites with these stations for its vast AWS customer base.

Conclusion

The race to dominate global logistics for two of the richest persons in the world is on. The media hypes up every story from these two entrepreneurial giants, but miss the overall aim that each has in the logistics industry. As much as Bezos seems to dominate the movement of parcels, Musk is slowly catching up with his approach to electric transport. For both Bezos and Musk, autonomous vehicles are the future of both human and freight transportation. Musk seems intent on using satellite Internet to operate autonomous vehicles on Earth. This conclusion logically follows from what the two men are currently doing in the logistics industry. To recap, let's look at what each is doing to revolutionize the logistics world, as covered throughout this book.

In Chapter 1, I covered the history of logistics and supply chain and revealed that both originated from military operations. Logistics is about the movement of goods and their storage, while supply chain includes moving raw materials, turning them into products, and distributing them. So, logistics can be thought of as a subset of a supply chain. To give a perspective, I next gave you an overview of both Jeff Bezos and Elon Musk as entrepreneurs.

The biggest challenge for Bezos' Amazon is to reduce shipping costs that equate to about or are less than 16% of its total sales. That way, this online retailer can increase its net profit and put a lot more money into shareholders' pockets. That's why Amazon's delivery and logistics strategy is crucial for it. Musk is purely interested in renewable energy and saving humanity. He has businesses in the automotive, telecommunications, energy, passenger transportation, healthcare, tunneling, and artificial intelligence spaces.

The rivalry between Bezos and Musk shows up clearly in the space race. From the late 1950s to around 2000, the space race pitted the United States and the then-Soviet Union. Things started changing when seven astronauts died when returning from the International Space Station via the Columbia shuttle. That's when Blue Origin and SpaceX started having opportunities to send their rockets and spaceships to space. SpaceX has dominated, winning numerous NASA and military contracts. The details of this space rivalry were shared in Chapter 2. SpaceX and Blue Origin sent manned vehicles to space in 2020 and 2021, respectively. However, SpaceX has gone to the International Space Station, 262 miles above earth, something that Blue Origin hasn't done yet.

In Chapter 3, I talked about the logistics race between Bezos and Musk. Amazon has a stronger logistics approach when compared with Tesla. This is made possible by having arms that focus on producing products and services geared to deliver customer packages faster. Amazon Robotics creates robots that simplify and improve warehouse and sortation centers' operational efficiencies. Through Amazon Logistics, Amazon has become a top four shipping provider, competing with FedEx and UPS, among other competitors. Tesla also has robots that help immensely in the production of its electric vehicles. However, a lack of car-haulage transportation crippled its objectives of delivering Model 3 cars to customers on time. This problem forced Tesla to innovate by building its own transport and trailers, introducing Tesla Direct, and by buying car-hauling trucks.

On the space front, Blue Origin has developed six engines for propelling its reusable rockets, which include the New Shepard and New Glenn. SpaceX developed the Falcon 9, the first reusable rocket in the world. It has also designed and produced the Falcon Heavy and Starship for launches to higher orbits, such as to the Moon and Mars. The Dragon is a spacecraft for conveying cargo and humans to space.

Bezos and Musk are also in the race for development of artificial intelligence (AI) as discussed in Chapter 4. This technology has become critical in the modern world, and can be used in various applications, including in autonomous cars and health care. Amazon's AI operates in its robots, as well as the online and offline selling of Amazon's products. Furthermore, AI plays a significant role in warehouse and sortation centers' operations to ensure customers receive their packages swiftly. Musk's companies utilizing AI include Tesla and Neuralink. Tesla uses AI to design autonomous vehicles. Hence, Tesla is now a technology company as much as it is an electric vehicle maker and energy company.

Another race involving Bezos and Musk is in the self-driving vehicle space. Tesla directly produces electric cars in both the luxury and mass market segments. It has also built its own self-driving software called Autopilot. However, no company has yet to produce a fully self-driving car or the technology for it. Currently, Tesla is in its tenth iteration of its Autopilot technology. You learned about this in Chapter 5. Through iterations of the self-driving technology, it's likely that Tesla and many other companies will produce full self-driving cars. Bezos' Amazon took a shortcut to taking part in the self-driving car space. It bought Zoox and invested in Aurora Innovation, two companies heavily involved in producing autonomous vehicles and car self-driving technology, respectively. This way, Amazon can move toward owning the bulk of its logistics and transportation needs.

Tesla had a head-start over Bezos and Amazon in the electric-vehicle market. It's already dominating the mass market, selling more than 500,000 cars in 2020 alone. Tesla designs and manufactures four models of electric vehicles: Model X, Model Y, Model S, and Model 3. More importantly, Tesla has four operational factories to ensure it makes enough cars for its target markets. All this information was covered in Chapter 6. In the years to come, Tesla will be busy producing the Cybertruck and Tesla Semi to compete in other vehicle segments. There's still the uncertainty of whether Tesla will make electric vans or not. Amazon invested in Rivian, an electric carmaker, and bought 100,000 vans for last-mile delivery. Furthermore, Amazon has ordered electric trucks from Lion Electric, a Canadian electric-vehicle maker.

In Chapter 7, I focused on the air-transportation race. Amazon already has Amazon Prime Air and Amazon

Air Cargo in this space. Both of these operations are involved in moving packages faster to the end customer. Prime Air is a drone-delivery service that has already been FCC-approved, while Amazon Air Cargo utilizes aircraft to deliver parcels between airports. Amazon buys and leases aircraft to make this package-delivery method successful. Tesla's air transportation has yet to take off. But, Musk is already dreaming about cargo continent-to-continent air travel using rockets, an initiative that has caught the interest of the U.S. Transportation Command. Furthermore, Musk is dreaming of developing a battery-operated supersonic jet. For now, Amazon seems to be winning the air-transportation race.

In Chapter 8, I shifted my attention to the freight-transportation race. Amazon operates an ocean freight business through its Chinese subsidiary and a freight-brokerage operation. Customers can ship containers from China to the United States by using Amazon Ocean Freight. Using these businesses, Amazon is able to control shipping costs and improve its bottom line. Tesla will be making the Tesla Semi, introduced in 2017, in the next few years. Several customers have already ordered the Tesla Semi, including Walmart and Pride Group Enterprises (PGE). Musk's The Boring Company is already pitching potential customers on freight tunnels. One of the tunnels accommodates a pair of shipping containers side by side in the tunnel and could improve shipping rates. Furthermore, Musk is considering developing the hyperloop for inter-city cargo transportation.

In Chapter 9, I shared with you the satellite-Internet-service race. SpaceX's Starlink is dominating this race since it has already deployed more than 1,000 satellites. Bezos' Project Kuiper is progressing and has recently completed testing its antenna to connect users with the Amazon network. Bezos and Musk tussled when SpaceX filed a request to the FCC to change its satellite orbit, a move Bezos and other companies don't approve. They believe this change could jeopardize their plans. Musk plans to use Starlink satellite Internet to operate autonomous vehicles.

The rivalry between Bezos and Musk looks like it won't end anytime soon, especially regarding space travel and satellite Internet. The figures below summarize the ultimate aim of Musk and Bezos as seen from a logical analysis of their activities. If there's one thing you should get, it is this: Both Bezos and Musk want to dominate the logistics on Earth. The difference is in how each wants to achieve this objective.

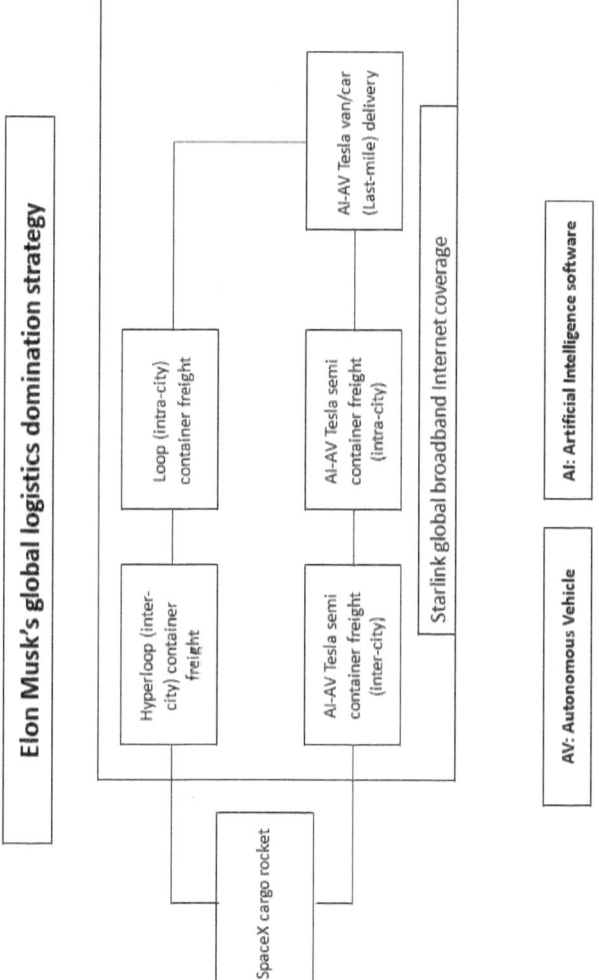

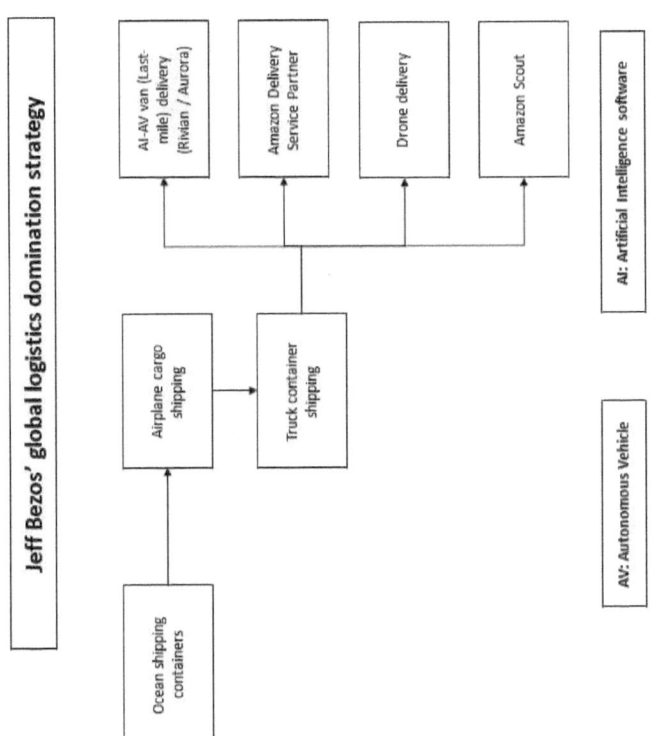

References

Amazon. (2019, September 19). *Amazon Co-Founds The Climate Pledge, Setting Goal to Meet the Paris Agreement 10 Years Early*. Amazon Press Center. https://press.aboutamazon.com/news-releases/news-release-details/amazon-co-founds-climate-pledge-setting-goal-meet-paris

Amazon. (2019, November 6). *Amazon Announces New Amazon Robotics Innovation Hub and Plans to Create 200 Tech and Advanced Manufacturing Jobs in Massachusetts*. Amazon. https://press.aboutamazon.com/news-releases/news-release-details/amazon-announces-new-amazon-robotics-innovation-hub-and-plans

Amazon. (2020, June 26). *We're Acquiring Zoox to Help Bring Their Vision of Autonomous Ride-Hailing to Reality*. Amazon. https://www.aboutamazon.com/news/company-news/were-acquiring-zoox-to-help-bring-their-vision-of-autonomous-ride-hailing-to-reality?ots=1&tag=curbedcom06-20&linkCode=w50

Amazon. (2021, July 29). *Amazon.com Announces Second Quarter Results*. Amazon. https://ir.aboutamazon.com/news-release/news-release-details/2021/Amazon.com-Announces-Second-Quarter-Results-2dcdc6a32/default.aspx

Amazon Freight. (n.d.). *Welcome to Amazon Freight*. Amazon Freight. https://freight.amazon.com/

Bellan, R., & Alamalhodaei, A. (2021, August 20). *Top Four Highlights of Elon Musk's Tesla AI Day*. TechCrunch. https://techcrunch.com/2021/08/19/top-five-highlights-of-elon-musks-tesla-ai-day/

Benjabutr, B. (2020, March 20). *History of Logistics and Supply Chain*. SupplyChainOpz. https://www.supplychainopz.com/2013/05/history-of-logistics.html

Blue Origin. (n.d.). *About Blue Origin*. Blue Origin. https://www.blueorigin.com/about-blue

Blue Origin. (n.d.-b). *Blue Engines: Rocket engines designed for reuse*. Blue Origin. https://www.blueorigin.com/engines/

Blystone, D. (2021, July 14). *Who is Elon Musk?* Investopedia. https://www.investopedia.com/articles/personal-finance/061015/how-elon-musk-became-elon-musk.asp

Bradsher, K. (2017, October 9). China Hastens The World Toward an Electric-car Future. *The New York Times*. https://www.nytimes.com/2017/10/09/business/china-hastens-the-world-toward-an-electric-car-future.html

Brown, M. (2019, June 7). *Tesla Electric Plane: Plans, Timeline For an Elon Musk Flying Machine*. Inverse.

https://www.inverse.com/article/57292-tesla-electric-plane-plans-and-release-dates-for-elon-musk-flying-machine

Business Standard. (n.d.). *Who is Elon Musk? Business Standard.* https://www.business-standard.com/about/who-is-elon-musk

Carlier, M. (2021, August 5). *Tesla Vehicles Delivered Between 1st Quarter 2020 And 2nd Quarter 2021, By Main Model.* Statista. https://www.statista.com/statistics/801157/tesla-quarterly-vehicle-deliveries-by-model/

Cava, M. della. (2017, November 17). *Tesla Semi, an Electric Big Rig Truck With 500-Mile Range, Rolls Into Reality.* USA TODAY. https://www.usatoday.com/story/tech/2017/11/16/tesla-semi-electric-big-rig-truck-rolls-into-reality/873162001/

Cheung, B. (2021, August 10). *Truck Driver Shortage "Is About As Bad As I've Ever Seen": US Xpress CEO.* Yahoo! Finance. https://finance.yahoo.com/news/us-xpress-ceo-driver-shortage-trucking-industry-202632404.html

Davis, D. (2020, May 26). *Amazon is the Fourth-Largest U.S. Delivery Service And Growing Fast.* Digital Commerce 360. https://www.digitalcommerce360.com/2020/05/26/amazon-fourth-largest-us-delivery-service/

Del Rey, J. (2019, December 11). *Amazon's Warehouse Robots and Their Complicated Impact on Workers.* Vox; Vox.

https://www.vox.com/recode/2019/12/11/209826 52/robots-amazon-warehouse-jobs-automation

Delbert, C. (2020, October 8). *How Strong are Elon Musk's 10-Cent bricks?* Popular Mechanics. https://www.popularmechanics.com/science/a3430 5094/elon-musk-boring-company-bricks-durability-test/

Duffy, K. (2021, April 27). *Jeff Bezos' Blue Origin Slams NASA for its "Unfair" Decision to Award Elon Musk's SpaceX a $2.9 Billion Contract.* Business Insider. https://www.businessinsider.com/jeff-bezos-blue-origin-elon-musk-spacex-nasa-contract-flawed-2021-4?IR=T

Federal Communications Commission. (2021, January 19). *Fourteenth Broadband Deployment Report.* Federal Communications Commission. https://docs.fcc.gov/public/attachments/FCC-21-18A1.pdf

Fox, E. (2021, March 1). *Tesla Autonomous Cars & SpaceX Starlink Will Likely Form a Necessary Synergy: Morgan Stanley offers analysis.* TESMANIAN. https://www.tesmanian.com/blogs/tesmanian-blog/morgan-stanley-on-tesla-autonomous-cars-and-starlink

Gallo, C. (2021, February 11). *How Jeff Bezos Consistently Communicates Four Core Values That Made Amazon a Success.* Forbes. https://www.forbes.com/sites/carminegallo/2021/ 02/11/how-jeff-bezos-consistently-communicates-four-core-values-that-made-amazon-a-success/?sh=137dcad76e24

Global Industry Analysts, Inc. (2021, June 25). *Global Freight Trucking Market to Reach $2.7 Trillion by 2026*. Cision PR Newswire. https://www.prnewswire.com/news-releases/global-freight-trucking-market-to-reach-2-7-trillion-by-2026--301320057.html

GreyB Services (2021, May 25). *Autonomous Vehicles Companies - A Market Research by GreyB Services*. GreyB. https://www.greyb.com/autonomous-vehicle-companies/

Haselton, T., & Bosa, D. (2019, February 7). *Amazon Just Invested in Self-Driving Car Company Aurora*. CNBC; CNBC. https://www.cnbc.com/2019/02/07/amazon-just-invested-in-self-driving-car-company-aurora.html

Isidore, C. (2021, January 6). *Tesla Short Sellers Lost $40 Billion in 2020. Elon Musk Made More Than Triple That*. CNN. https://edition.cnn.com/2021/01/06/investing/tesla-shorts-losses-elon-musk-win/index.html

Juliussen, E. (2021, March 11). *Amazon Quietly Worming its Way into the Auto Industry*. EETimes. https://www.eetimes.com/amazon-quietly-worming-its-way-into-the-auto-industry/#

Kingston, J. (2021, February 10). *Uber Freight's Q4 Numbers Up From 2019 and From Third Quarter*. FreightWaves.

https://www.freightwaves.com/news/uber-freights-q4-numbers-up-from-2019-and-from-third-quarter

Kolakowski, M. (2019, September 19). *What is Rivian?* Investopedia. https://www.investopedia.com/what-is-rivian-4692084

Lambert, F. (2019, March 11). *Tesla is Buying Car-Hauling Trucks and Trailers Using $13 Million in TSLA Shares.* Electrek. https://electrek.co/2019/03/11/tesla-buying-trucks-trailers-tsla-shares/

Louli, A. J., Eldesoky, A., Weber, R., Genovese, M., Coon, M., deGooyer, J., Deng, Z., White, R. T., Lee, J., Rodgers, T., Petibon, R., Hy, S., Cheng, S. J. H., & Dahn, J. R. (2020). Diagnosing and Correcting Anode-Free Cell Failure via Electrolyte and Morphological Analysis. *Nature Energy, 5*(9), 693–702. https://doi.org/10.1038/s41560-020-0668-8

Mahnken, D. (2018, October 6). *What is Logistics? Origins, Present and Future Scenarios.* Saloodo!Blog. https://www.saloodo.com/blog/logistics-meaning/

Manthey, N. (2020, September 30). *Walmart Canada triples its Tesla Semi order.* Electrive.com. https://www.electrive.com/2020/09/30/walmart-canada-orders-another-100-tesla-semi/

McBride, S. (2021, June 23). *Elon Musk's Boring Company Pitching Wider Freight Tunnels for Containers.* GCaptain. https://gcaptain.com/elon-musks-boring-company-pitching-wider-freight-tunnels-for-containers/

Mordor Intelligence. (n.d.). *Electric Vehicle Market Report (2021 – 2026) — COVID Impact, Size, Share, Industry*

Growth. Mordor Intelligence. https://www.mordorintelligence.com/industry-reports/electric-vehicle-market

Müller, W. (1811). The Elements of The Science of War: Containing The Modern, Established, and Approved Principles of The Theory and Practice of The Military Sciences. In *Google Books*. Longman, Hurst, Rees, Orme and Company. https://play.google.com/books/reader?id=NcfUpuWtuNEC&pg=GBS.PP7&hl=en

Musk, E. (2016, July 21). *Master Plan, Part Deux*. Tesla. https://www.tesla.com/blog/master-plan-part-deux

Musk, E. (2018, November 19). *Maybe interesting to work with Daimler/Mercedes on an electric sprinter. That's a great van. We will inquire.* Twitter. https://twitter.com/elonmusk/status/1064541185820114944

Musk, E. (2020, October 25). *Sigh ... There should be a new supersonic jet, this time electric.* Twitter. https://twitter.com/elonmusk/status/1320180525009571845

Musk, E (2021, January 26). *It does not serve the public to hamstring Starlink today for an Amazon satellite system that is at best several.* Twitter. https://twitter.com/elonmusk/status/1354018055014260738

Neuralink. (n.d.). *About*. Neuralink. https://neuralink.com/about/

O'Callaghan, J. (2020, October 27). *SpaceX Reveals Monthly Cost of Starlink Internet in its "Better Than Nothing Beta."* Forbes. https://www.forbes.com/sites/jonathanocallaghan/2020/10/27/spacex-reveals-monthly-cost-of-starlink-Internet-in-its-better-than-nothing-betabut-is-it-too-expensive/?sh=377cf5a04a83

Online Etymology Dictionary. (n.d.). *Origin and Meaning of Logistics.* Online Etymology Dictionary. https://www.etymonline.com/word/logistics

OpenAI. (2015, December 11). *Introducing OpenAI.* OpenAI. https://openai.com/blog/introducing-openai/

OpenAI. (n.d.). *OpenAI charter.* OpenAI. https://openai.com/charter/

Palmer, D. (2018, March 22). *What is the Tesla Semi? Everything You Need to Know About Tesla's Semi-Autonomous Electric Truck.* ZDNet; ZDNet. https://www.zdnet.com/article/what-is-the-tesla-semi-everything-you-need-to-know-about-teslas-semi-autonomous-electric-truck/

Plainsite. (2021, May 3). *California DMV Tesla Robo-Taxi / FSD e-mails.* Plainsite. https://www.plainsite.org/documents/242a2g/california-dmv-tesla-robotaxi--fsd-emails/

Prince, T. (2020, April 2). *Word of the Year: Supply Chain.* Journal of Commerce Online. https://www.joc.com/international-logistics/word-year-supply-chain_20200402.html

Reichert, C. (2021, February 4). *Elon Musk's SpaceX Starlink Internet Service Now Has 10,000 Users.* CNET. https://www.cnet.com/home/Internet/elon-musks-spacex-starlink-Internet-service-now-has-10000-users/

Reuters Staff. (2020, April 15). U.S. Self-Driving Car Startup Zoox Agrees to Settle Lawsuit With Tesla. *Reuters.* https://www.reuters.com/article/us-tesla-lawsuit-zoox-idUSKCN21X034

Sage, A., & Taylor, E. (2018, May 25). Exclusive: Tesla Flies in New Battery Production Line for Gigafactory. *Reuters.* https://www.reuters.com/article/us-tesla-gigafactory-production-exclusiv-idUSKCN1IQ2RN

ScaleyMurrsuit. (2018, November 19). *Shame that's not a Tesla VAN. Because we REALLY REALLY REALLY REALLY need a good tough as nails electric trades van.* Twitter. https://twitter.com/ScaleyMurrsuit/status/1064540665210519552

Seeking Alpha. (2021, April 27). *Tesla, Inc.'s (TSLA) CEO Elon Musk on Q1 2021 Results – Earnings Call Transcript.* Seeking Alpha. https://seekingalpha.com/article/4421439-tesla-inc-s-tsla-ceo-elon-musk-on-q1-2021-results-earnings-call-transcript

Shaban, H. (2018, October 2). *Tesla Meets its Model 3 Production Goals for Third Quarter.* Washington Post. https://www.washingtonpost.com/technology/201

8/10/02/tesla-meets-its-model-production-goals-third-quarter

Sheetz, M. (2021, January 26). *Elon Musk Blasts Jeff Bezos' Amazon, Alleging Effort to "Hamstring" SpaceX's Starlink Satellite Internet.* CNBC. https://www.cnbc.com/2021/01/26/elon-musk-blasts-jeff-bezos-amazon-competitor-to-spacexs-starlink-.html

SpaceX. (n.d.). *Falcon 9: First orbital class rocket capable of reflight.* SpaceX. https://www.spacex.com/vehicles/falcon-9/

SpaceX. (n.d.). *Falcon Heavy: The world's most powerful rocket.* SpaceX. https://www.spacex.com/vehicles/falcon-heavy/

SpaceX. (n.d.). *SpaceX.* SpaceX. https://www.spacex.com/mission/

SpaceX. (n.d.). *Starship.* SpaceX. https://www.spacex.com/vehicles/starship/

Starlink. (n.d.). *Starlink.* Starlink. https://www.starlink.com/

Tesla. (n.d.). *Solarglass roof.* Tesla. https://www.tesla.com/solarroof

Tesla. (n.d.). *Tesla Semi.* Tesla. https://www.tesla.com/en_CA/semi

Tesla, Inc. (2021). *Q2 2021 update.* Sec.gov. https://www.sec.gov/Archives/edgar/data/1318605/000156459021037953/tsla-ex991_89.htm

The Boring Company. (n.d.). *Prufrock*. The Boring Company. https://www.boringcompany.com/prufrock

The Lion Electric Co. (2020, September 23). *Lion Electric to Deliver 10 All-Electric Trucks to Amazon*. Www.prnewswire.com. https://www.prnewswire.com/news-releases/lion-electric-to-deliver-10-all-electric-trucks-to-amazon-301136284.html

Vetter, D. (2020, November 18). *Britain Will Ban Gasoline Cars in 2030. Why Are Experts Not Impressed?* Forbes. https://www.forbes.com/sites/davidrvetter/2020/11/18/britain-will-ban-gasoline-cars-in-2030-why-are-experts-not-impressed/?sh=6a82107636e4

xChange. (2019, October 7). *The Amazon NVOCC Disruption*. XChange. https://container-xchange.com/blog/the-amazon-nvocc-disruption/

www.ingramcontent.com/pod-product-compliance
Lightning Source LLC
Chambersburg PA
CBHW030908080526
44589CB00010B/209